Good Design Is Good Business China Awards

GOOD DESIGN IN CHINA

中国好设计

（美）克里福德·皮尔逊 编　韩雪婷 译

辽宁科学技术出版社

图书在版编目（CIP）数据

中国好设计／（美）克里福德·皮尔逊编；韩雪婷译. —— 沈阳：辽宁科学技术出版社，2011.2

ISBN 978-7-5381-6664-4

Ⅰ．①中… Ⅱ．①克… ②韩… Ⅲ．①建筑设计－作品集－中国－现代 Ⅳ．①TU206

中国版本图书馆CIP数据核字（2010）第179109号

出版发行：辽宁科学技术出版社

（地址：沈阳市和平区十一纬路29号　邮编：110003）

印　刷　者：利丰雅高印刷（深圳）有限公司

经　销　者：各地新华书店

幅面尺寸：230mm × 300mm

印　　张：41

插　　页：4

字　　数：50千字

印　　数：1~2000

出版时间：2011年 2 月第 1 版

印刷时间：2011年 2 月第 1 次印刷

责任编辑：陈慈良　韩雪婷

封面设计：曹　玲

版式设计：曹　玲

责任校对：周　文

书　　号：ISBN 978-7-5381-6664-4

定　　价：258.00元

联系电话：024-23284360

邮购热线：024-23284502

E-mail: lnkjc@126.com

http://www.lnkj.com.cn

本书网址：www.lnkj.cn/uri.sh/6664

Contents 目录

From Highrise to Hutong Bubble: The Growing Role of Design in China Today

从高层到胡同泡泡：中国今日设计的角色成长

By Clifford A. Pearson
Deputy Editor, Architectural Record

文 克里福德·皮尔逊
《建筑实录》责任编辑

Whenever I speak with architects about their most successful projects, I hear them talk about the importance of collaboration. The best architects create collaborative environments in their offices so that everyone on their team can contribute ideas, passion, and expertise. But they also emphasize the need for collaboration with other designers – such as engineers, landscape architects, and lighting consultants – and especially with clients.

True collaboration is often difficult to achieve. It is always easier for one architect or one firm to make all the key decisions and not worry about what anyone else might want or suggest. But such autocratic design rarely provides the richness of vision that comes when people from different groups bring a broad range of perspectives to bear on a particular project. Listening, responding, compromising, and rethinking take extra time, but they increase the chance that innovation will take root in a project's design.

The most difficult part of listening and collaborating is truly understanding the other person's needs. When working with a client, a good architect immerses himself (or herself) in that person's business and learns everything he can about it. He learns about the client's products and services, his client's competition, and his client's business strategy. Only then can he earn the client's trust and take the client in new directions. Only then can he propose design ideas that respond to the client's underlying needs and push that company forward. Once a client realizes that you know as much about his business as he does, he will listen to your ideas and consider the design strategies that he might have dismissed before.

Such collaboration has driven all the winners of Architectural Record's Good Design Is Good Business Awards since the magazine launched the program in the United States in 1997 and in China in 2006. Reflecting the spirit of the awards, Record collaborated with BusinessWeek magazine, which was also published by our parent company, McGraw-Hill. (BusinessWeek was sold to Bloomberg LP in 2009.) Most architectural awards programs focus just on design, but our program looks at the intersection of design and business. Judges evaluate a project not only on how it looks, but how it performs. Entrants have to show how their project furthers the goals of their client– whether that client is a corporation, a cultural organization, a government agency, or an educational institution.

Getting accurate measurements – "metrics", as business people say – of performance is not always easy. You can quantify things like energy use and water recycling, but it is harder to put a number on the effect of a new building on a company's ability to attract the best workers and retain them. What impact does architecture have on worker productivity and worker satisfaction? How does a building change a company's image among the public and its success in marketing its goods and services? Such things are all very difficult to quantify. But the Good Design Is Good Business Awards challenges entrants to do so. Although often imprecise or vague, entries must include an explanation of the impact of design on performance.

Instead of honoring architects, the Good Design Is Good Business Awards program honors the building team, which includes the client, the architect, and consultants. Winners are business people and organizational leaders, as well as designers. As a result, the program is different from almost all other design awards. As Mack Scogin, a partner in the Atlanta-based firm Mack Scogin Merrill Elam Architects and a former chairman of the department of architecture at Harvard's Graduate School of Design, explained when the program began more than a decade ago, "It has the potential of being an incredibly important awards program. It gets at what architects can do. It's all about challenging clients. It's about how architects can affect a client's need."

In China, we organize the awards every second year, so we recently completed our third cycle. In that time, we have honored 45 projects in a range of categories (best public project, best commercial project, best residential project, best planning project, best green project, best preservation project, and occasionally, best interior). Each cycle, we also feature one "Best Client," honoring a company or agency that uses architecture as a critical element in its mission and strategy. In 2010, we added extra excitement by naming one Grand Winner in each project category and a Project of the Year.

Looking back at all these winners, I see a wonderful range of buildings – from single-family houses to highrise apartment towers, from corporate headquarters to university buildings, from preservation guidelines for a historic town in Yunan to master plans for great metropolitan districts. This diversity says a lot about what is happening in China today where important construction is moving forward in big cities and rural villages.

Similarly, winning architects come from China and abroad, from big firms and small ones. Winning clients run large development companies, university departments, and government agencies. But all of them believe in the power of design to affect change and produce innovation. Some of this behavior was based on a leap of faith– trusting an architect to do the right thing–but much of it was based on a pragmatic, no-nonsense examination of facts and figures. Increasingly, architects and clients are learning how to measure the impact of design on the business bottom line.

Viewed together, the 45 winning projects and three best clients provide a fascinating snapshot of China in the 21st century. I look at all the museums that have won awards including the Dafen Art Museum, the Liangzhu Museum, the Suzhou Museum, the Tangshan Urban Planning Museum and Park, the Shanghai Xiang-Dong Buddhist Art Museum, and the Luyeyuan Stone Sculpture Museum – and I see a cultural renaissance that has swept over the country. During the past 10 years, contemporary Chinese art has soared in value on the international market and is now proudly exhibited in New York, Los Angeles, Paris, Berlin, London, Sydney, and Singapore, as well as Beijing, Shanghai, and Chongqing. This trend tells me that China is now a place where innovation in art has become part of the national DNA. Since one discipline often influences others, I am sure cutting-edge art will encourage new ways of thinking in business, management, architecture, and other fields. In a world connected by electronic media, breakthroughs in art and culture inevitably spark imaginations in the business world as well.

I also see a growing reliance on innovative planning. Projects such as the master plan for the Olympic Green in Beijing and the plan for Meixi Lake, an entire new city adjacent to Changsha, tell me that both developers and government agencies are undertaking sophisticated projects that weave together mass transit, mixed uses, and pedestrian-oriented public spaces. Such planning expertise, though, is also being applied to smaller towns such as Longchi in Sichuan Province and Qiaonan Village in Fujian Province. So good design is not only good urban planning, it is also good rural planning.

Another important trend is the rise of green design in all kinds of building. While China's three-decade-long economic boom has created huge environmental problems (according to Forbes magazine, the 10 most polluted cities in the world are all in China), the nation's leaders are now directing enormous resources to fixing the problem and pioneering environmentally sustainable technologies. So China has already become the world's leader in the production of wind turbines and solar panels and is connecting its cities by high-speed rail lines that will reduce the need for people to travel by automobile. Winners in our green project category – including the recent IBR Headquarters and Vanke Center, both in Shenzhen – are some of the most environmentally responsible buildings in the world and are models that architects from other countries can learn from. I think clients and designers from the United States, Europe, Latin America, the Middle East, and Africa will increasingly look to China for answers to the problems of energy consumption, climate change, and pollution.

Reviewing all the winners of the Good Design Is Good Business Awards, I also see the emergence of a talented generation of Chinese architects and clients. Firms such as Urbanus, MADA s.p.a.m., Jiakun Architect & Associates, Atelier Feichanglianzhu, MAD Architects, and Atelier Deshaus are producing innovative projects that are changing the way architects everywhere think about design. And clients such as Shui On Land Ltd, China Vanke Co. Ltd, and Shanghai Qingpu New Town Construction Company are establishing themselves as models for using architecture to further their business and organizational goals.

Because China's economy is pushing forward as many other nations falter, more and more of the best architects from around the world will be working on projects in China. As a result, many of the most important buildings will rise here, contributing to a culture of architectural innovation. While I'm sure there will be bumps along the road – as there are in every country – the great arc of economic progress will most probably continue for quite a while and make China the "Middle Kingdom" once again.

The great response we have received from our readers about the Good Design Is Good Business China Awards encourages us to push forward with the program and hopefully expand it. Working with our partners at LNSTP and Time + Architecture magazine, we are optimistic about using the awards program to improve the performance of architects and clients throughout China and to inform the conversation about design's role in society in general.

[signature]

每当我与建筑师谈论他们最成功的作品，我总是听到他们强调合作的重要性。最好的建筑师会在他们的工作室内部建立起团队合作的风气，使得队伍中的每个人都可以贡献出创意、激情以及专门的技术。但是，他们同时也非常注重与其他设计师的合作——如工程师、景观建筑师、照明系统设计师等——尤其是看重与他们的客户良好配合。

真正的合作往往很难达成。对一位建筑师或设计公司来说，独自做出关键性的决定而不需考虑别人的想法或建议比较容易。但是这样"独裁"的项目很难像那些采纳了不同方面的意见及想法的设计一样达到丰富的视觉效果。听取、反映、妥协、重新考虑，这些过程都要花费额外的时间，但是这样也会增加创新意识在作品中扎根的机会。

听取意见及做出合作姿态的最困难之处就在于如何真正地了解对方的需要。在为客户工作时，一位好的建筑师会让他自己（她自己）设身处地的站在业主的角度上，尽可能地感知业主的一切要求。他要了解客户的产品及服务，竞争对手情况及经营的策略。只有这样，他们才可以赢得客户的信任，指导客户在新领域中的发展。这样设计师才可以提出适应客户各项潜在需求的设计并且推动整个公司的发展。一旦客户意识到设计师对他的业务的了解如同他自己一样多，他会听从设计师的想法，并重新考虑某些也许之前他反对过的意见。

1997年在美国及2006年在中国发行的《建筑实录》杂志中，评出的"好设计创造好效益"奖项的获奖者，都是这种良好合作的实例。为了体现这个奖项的精神，《建筑实录》也与同是总公司麦格劳希尔旗下的出版物《商业周刊》进行了合作（《商业周刊》2009年被卖与Bloomberg LP）。大多数的建筑奖都是关注于设计本身，而我们则更为注重设计与商业效益的结合。评审团队不仅要强调项目的外观，更重要的是考察项目实际的运用效果。参选者必须展现出他们的设计如何协助业主达成事业目标——无论业主是一个公司、一个文化组织、一个政府机构或者是一个教育机构。

得到测评项目商业表现的准确指标——商人们常说的"定量"——非常困难。你可以准确得知能源使用或水量循环的数值，但是对于一个公司新建的大楼对其优秀员工的吸引和存留能力的考量，却无法用某一个数据去表示。一个建筑在员工的生产力和满意度上有哪些影响？一座大厦如何改变一个公司在公众眼中的印象，如何促使公司在产品和服务上获得成功？这种指标很难被量化。但是好设计奖却在冲击这种挑战，虽然有时也不够精确或表达隐晦，但参选作品必须对其设计对实际运用效果的影响力做出解释。

"好设计创造好效益"奖，是颁发给整个建筑团体的荣誉——包括业主、设计师及建筑顾问等，

而不单单是褒奖设计师个人。获奖者是商人、组织领导，还有设计师。因此，这个奖项可以区别于其他所有设计类奖项。亚特兰大Mack Scogin Merrill Elam 建筑师事务所合伙人、哈佛大学设计学院建筑系前主席Mack Scogin在10多年前开始评奖活动时就曾解释说："好设计奖有潜力成为未来最重要的大奖之一，它关注设计师的能力和对业主的挑战，它表现了设计师如何满足了业主的需要。"

在中国，我们每两年举行一次评奖，因此现在刚刚进行到了第三届。在此期间，共有45个项目获得不同类别的大奖（最佳公共建筑奖、最佳商业建筑奖、最佳住宅项目奖、最佳规划项目奖、最佳绿色项目奖、最佳历史保护项目奖及个别的最佳室内奖）。每一届，我们都选出一个"最佳业主"，以表彰某个公司或组织将建筑作为自己的经营目标和战略中的关键性要素。在2010年，我们还为好设计奖提供了额外的惊喜——每个类别的奖项中都评出了一个杰出项目大奖和一个年度最佳项目大奖。

回顾这些获奖作品，我看到了一组美妙的建筑作品——从独幢的家庭别墅到高层的商务住宅，从集团总部到校区教学楼，从云南历史古镇的保护方案到大都市的发展规划。这种跨度体现了当今中国从大城市到小乡村都在推进的伟大建设的变化。

同样，获奖建筑有中国的，也有海外的，有的来自大型事务所，有的来自小型工作室，获奖业主有大型发展公司，也有学校院系，还有政府机构。但是，他们都相信设计可以带来改变及创新。有时这种合作建立在突发的信任基础之上——相信建筑师在做最合适的设计——然而大部分的合作达成还是依靠务实的、严肃的对事实及数据的考察而得来。建筑师和业主都在越来越多地学会考量设计在商业底线的巨大影响。

总体观之，这45个获奖项目及三个最佳业主大奖，为21世纪的中国展现了一幅引人入胜的画卷。我去看过所有获奖的博物馆——包括大芬美术馆、良渚博物馆、苏州博物馆、唐山城市展览馆、上海相东佛像馆、鹿野苑石刻博物馆——在这里我看到了遍布中国的文化新生。在过去的10年里，中国当代艺术在国际市场上价值飞升，以傲人的姿态在纽约、洛杉矶、巴黎、柏林、伦敦、悉尼、新加坡以及北京、上海、重庆等地进行了展览。这种趋势告诉我们，中国现在已经成为了以艺术创新为民族核心的国度。由于一种学科往往会影响到其他学科的发展，我们相信，尖端的艺术也会鼓励起在商业、管理、建筑和其他领域的全新的思维方式。在这样一个电子媒体发达的世界，艺术的突破也不可避免地为商业世界带来新思维的爆发。

同时，我也看到了对创新规划的依赖日益增加。一些项目，比如北京奥林匹克公园总规划和长沙附近的一个新城——梅溪湖规划，让我感受到开发商和政府承担起了一种集合了大规模运输、多功能运用以及公共空间内容的经典项目。这样的专门规划技巧，同时，也应用在了一些乡村项目，比如四川省的龙池镇规划和福建省的桥南村改造上。可以说，好设计奖不仅是都市的好设计，也是乡村的好设计。

另一个重要的趋势是，在所有种类的建筑中，绿色设计都被强调为重中之重。中国30年的经济腾飞造成了巨大的环境问题（据福布斯数据，世界上10大污染城市全部集中在中国），现在国家领导人正在调动极大的资源来解决这个问题并积极寻求环境可持续发展的方法。所以现在中国已经成为了世界风动机械和太阳能电池产品的先锋国家，并且建设更多的城际高铁来减少人们出行时对汽车的需要。我们奖项中的最佳绿色项目奖得主——包括同在深圳的最新获奖的深圳建科大楼和万科中心——在世界范围内也是最具有环境责任意识的建筑，是其他国家建筑师可以学习的典范。我认为，美国、欧洲、拉美、中东及非洲的业主和设计师都可以从中国得到能源消耗、气候变化和环境污染等问题的答案。

回顾"好设计创造好效益"所有奖项得主，我看到了中国新一代天才建筑师和业主的诞生。一些设计事务所如都市实践、马达思班、家琨建筑、非常建筑、马岩松事务所、大舍建筑等的创新项目正在改变建筑师原有的设计思维方式。一些业主，如瑞安房地产发展有限公司、中国万科及上海青浦新城建设发展公司等也成为了运用建筑进一步实现企业和机构目标的模范。

由于在众多国家经济低迷期间，中国依然保持经济增长，越来越多的世界优秀建筑师会在中国开发他们的新项目。其结果是，越来越多的重要建筑在这里拔地而起，形成新的建筑文化。前进的路上会有诸多困难——如同在任何国家——但经济发展的弧线会长时间延续并使中国再次成为"中部王国"。

我们从《中国好设计》的读者群中获得的良好反馈将鼓励我们继续推进这项活动并争取将这个奖项做大。在与我们的合作伙伴辽宁科学技术出版社和《时代建筑》杂志的工作过程中，我们对利用这些获奖项目来提升中国当代建筑师和业主的影响力及表现力，向世界诠释设计师在社会中所扮演的重要角色，持有非常乐观的态度。

Chervon International Trading Company

泉峰国际贸易公司

Traditional Chinese garden design influenced the building configuration and landscaped spaces of the corporate headquarters for Chervon, a Chinese exporter of power tools. Located in the Nanjing Economic Development District on the outskirts of the city, the 30,700-square-meter building houses five major corporate departments in its five wings: management, sales, research and development, testing and training.

Taking its inspiration from the traditional zigzag garden path, the building bends to form two exterior spaces, the entrance court, which is open on its eastern edge and a more private west-facing garden. A narrow circulation spine cuts north/south across the site linking the five building wings and a series of courtyards. It alternately passes through the building wings and the courtyards, at times bridging over water like the traditional garden path. The two types of paths present in the building suggest the company's two cultural aspects, one represented by the non-direct contemplative zigzag and the other by the direct, practical line. The organization of courtyards linked on an axis has precedent in Chinese monastery design.

The roof forms one continuous slope from north to south, from the sixth story to the second, and from the public court to more intimate garden. In places it floats above the building as a trellis or overhang, whereas at others it seems to be part of the solid stone-clad block of the building. The roof itself is a landscaped green roof.

The focal point of the entry sequence is the asymmetrical pyramid that pierces the two-story lobby. Clad in aluminum of varying tooled finishes, it symbolizes the company's product and is intended as a space to display examples of Chervon's tools.

传统的中国园林式设计影响了泉峰集团——中国一家电动工具出口商——总部大楼的建筑规划及景观空间。此项目位于南京市郊的经济开发区，占地30,700平方米，共有五大分区，配备给集团的五大部门：管理、销售、研究与开发、检测及培训。

受到传统的"之"字形园林小径的启发，建筑采取了弯曲的造型，形成两个外部空间，有大厦的入口前庭，东面开放，另有一个朝西的较私密的花园。一条狭窄的中心隆起地点，横跨南北，连接了该建筑的五大主体部分和一些周围的庭院。它穿过大楼的各个部分和院落，有时也充当水面上的小桥，其作用与传统园林里的小径相似。两种不同风格的路径暗示出泉峰集团两种不同的文化战略，一种是非直接指引的"之"字形小路，另一种是直接实用的路径。由中轴统领院落组织的设计曾经在中国寺院的修建中有先例。

屋顶形成由北至南的倾斜，由六层过渡至二层，由开放庭院转为私人花园。在一些地方，屋顶在大楼之上形成格架或突起，在另一些地方似乎成为了大楼外表面石块的一部分。可以说，屋顶本身就是一种景致美化了的绿色屋顶。

大楼入口的亮点是穿过二层大厅的不对称的金字塔，铝制外墙的变幻象征着集团的产品，并且意图提供展示泉峰集团工具的空间。

Project name: Chervon International Trading Company
Award date: 2010
Location: Nanjing, China
Building area: 30,700 m²
Architect: Perkins+Will Inc., Architectural Design & Research Institute of Southeast University
Client: Chervon Group
Photographer: James Steinkamp of Steinkamp Photography
Completion date: 2007
Award name: McGraw-Hill Construction 3rd Bi-Annual "Good Design Is Good Business" China Awards 2010, Best Commercial Project

项目名称：泉峰国际贸易公司
获奖时间：2010
项目位置：中国 南京
建筑面积：30,700平方米
建筑设计：帕金斯威尔建筑师事务所   东南大学建筑设计研究院
业主：泉峰集团
摄影师：James Steinkamp of Steinkamp Photography
完成时间：2007
所获奖项：麦格劳-希尔公司《建筑实录》、《商业周刊》第三届"好设计创造好效益"中国奖项 2010最佳商业建筑

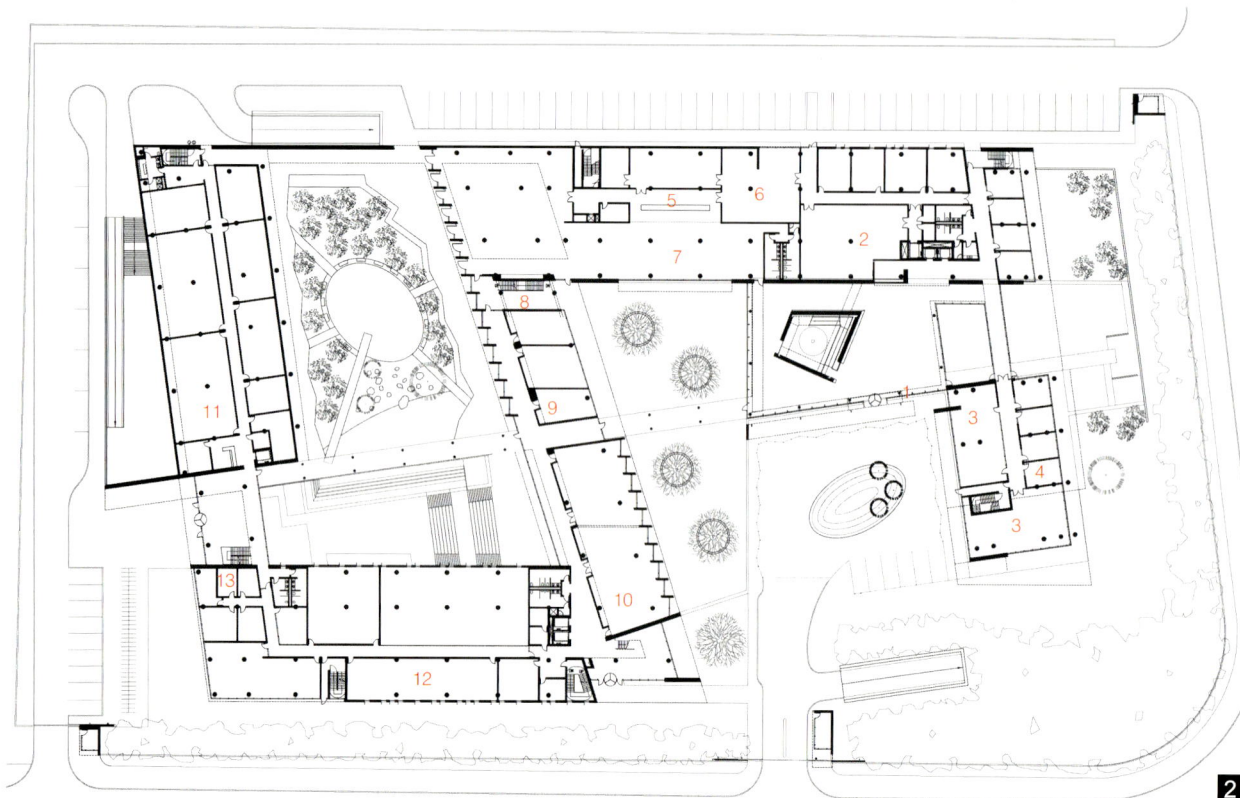

1. Main Lobby
   正厅
2. Storage & Receiving
   收发室
3. Showroom
   展厅
4. Meeting Rooms
   会议室
5. Kitchen
   厨房
6. Dishwashing
   清洗室
7. Servery
   备餐室
8. Employee Dining
   职工餐厅
9. Training Rooms
   培训室
10. Multipurpose Room
    多功能厅
11. Testing Labs
    实验室
12. Research & Development
    Open Office
    研发办公室
13. Lobby
    大厅

3

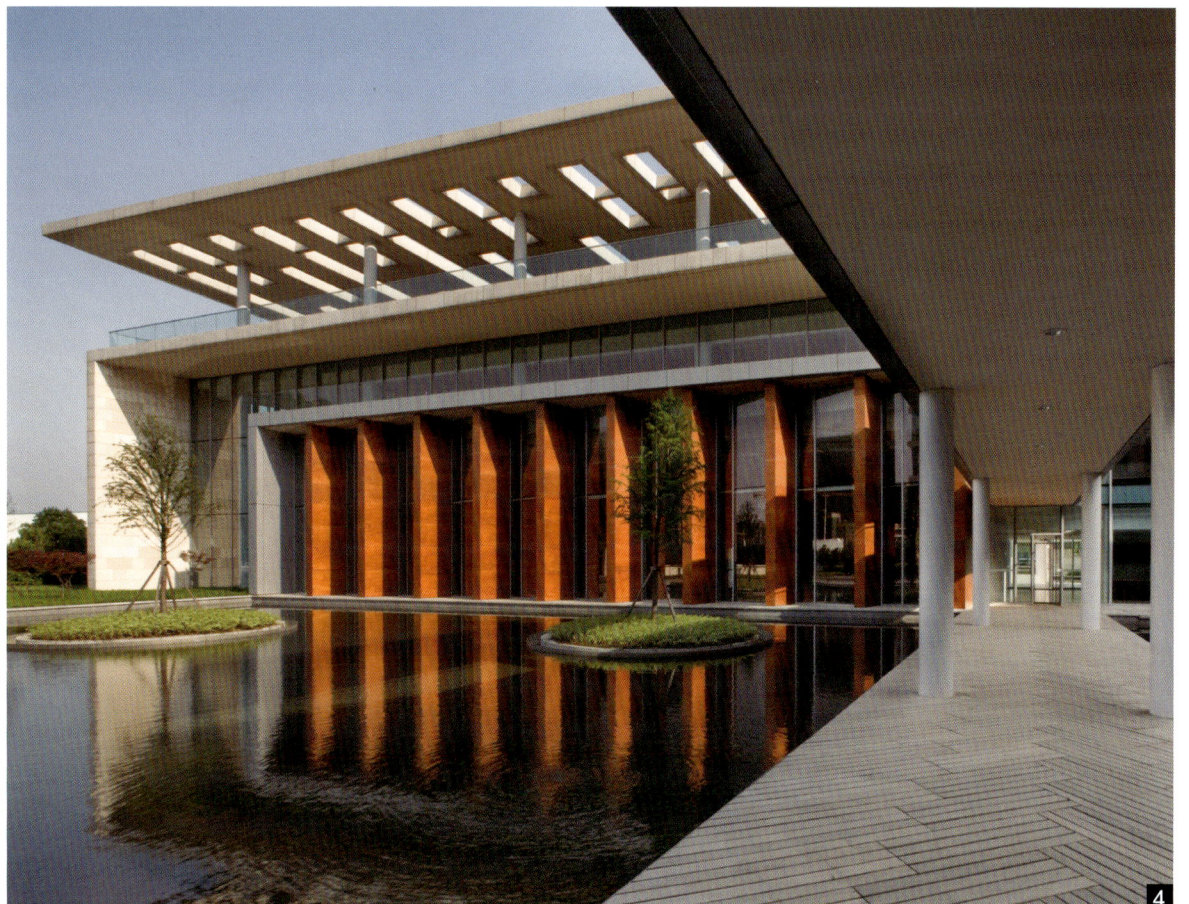

4

1. Aerial View from Southwest
   东南方向鸟瞰图
2. Plan
   平面图
3. Training Wing from Public Courtyard
   公共区培训部
4. Training Wing and Circulation Spine
   培训部及循环系统

1. Entry Pavilion with Pyramid, from Training Wing

   培训部金字塔形的入口

2. Bridge to Management Wing

   管理部的通道桥

3. Private Courtyard from Guest Room in Training Wing

   管理部待客区的私人空间

1. Entry Pavilion with Pyramid
   金字塔形入口

2. Water Feature in Private
   Courtyard
   内院中的水景

3. Rooftop Garden and Running
   Track above Training Wing
   培训部顶楼花园及步道

4. Entry Pavilion with Pyramid,
   from Bridge to Management
   Wing
   入口及过道桥

Shanghai World Financial Center
上海环球金融中心

A square prism – the symbol used by the ancient Chinese to represent the earth – is intersected by two cosmic arcs, representing the heavens, as the tower ascends in gesture to the sky. The interaction between these two realms gives rise to the building's form, carving a square sky portal at the top of the tower that lends balance to the structure and links the two opposing elements – the heavens and the earth.

A virtual city within a city, the 381,600-gross-square-meter SWFC houses a mix of office and retail uses, as well as a Park Hyatt Hotel on the 79th to 93rd floors. Occupying the tower's uppermost floors, the SWFC Sky Arena offers visitors aerial views of the historic Lujiazui and winding river below and the chance to literally walk almost 500 meters above the city via the 100th-floor Sky Walk. A large retail volume wraps around the base of the tower and faces a planned public park on the site's eastern side, further activating the sphere of activity at street level.

The elemental forms of the heavens and the earth are used again in the design of the building's podium where an angled wall representing the horizon cuts through the overlapping circle and square shapes. The wall's angle creates a prominent facade for the landscaped public space on the tower's western side, and organizes the ground level to provide separate entrances for office workers, hotel guests and public access to express elevator service for Sky Walk visitors.

Originally conceived in 1993, the project was put on hold during the Asian financial crisis of the late 1990s and was later redesigned to its current height – 32 meters higher than previous. The new, taller structure would not only have to be made lighter, but would need to resist higher wind loads and utilize existing foundations which had been constructed prior to the project delay. The project's structural engineer, Leslie E. Robertson Associates, arrived at an innovative structural solution which abandoned the original concrete frame structure in favor of a diagonal-braced frame with outrigger trusses coupled to the columns of the mega-structure. This enabled the weight of the building to be reduced by more than 10%, consequently reducing the use of materials and resulting in a more transparent structure in visual and conceptual harmony with the tower's elegant form.

古代中国讲究天圆地方，建筑的主体采用了象征大地的正方形柱体，两侧大器的弧线象征天空，在正方形柱体上形成两个拱形切面，伸向天际。两种元素的交叠使建筑的形态优雅而伟岸。塔顶还打造了一个空中门户，以平衡建筑结构，并将天与地两个对立的元素连接起来。

上海环球金融中心可以说是城中之城，总面积为381,600平方米的塔楼中集合了办公、商业以及79至93层的柏悦酒店。空中观光大厅位于塔楼顶部，游客在那里可以俯瞰著名的陆家嘴金融区和蜿蜒曲折的黄浦江，还可以在100层的观光厅体验在城市上空近500米的高度自由行走。塔楼的底部有商业环绕，东侧面朝公园，进一步提升了底层的活力。

建筑在裙楼设计上再次使用了天空、大地这两个元素形式，象征地平线的直角墙穿越交叠的圆形和方形结构。墙在折角处的立面设计最为突出，在塔楼西侧形成了一个设有景观的公共空间，并对地面层的布局进行了有效组织，为办公人员、酒店客人以及前往观光大厅的游客提供了单独的出入口。

项目在1993年提出了最初的设想，20世纪90年代遭遇亚洲金融危机而被搁置，最终得以重新设计，其高度在原设计上增加了32米，才有了现在的环球金融中心。新设计更高，面对的挑战不仅在于整体重量需要减轻，还需经受更强风力的考验，并且只能以工程停滞前所建的地基为基础。项目的结构工程师 —— Leslie E. Robertson联合公司，放弃了原有的混凝土框架结构，而采用了创新的结构解决方案，以斜撑框架与外伸桁架与主结构连接。这样一来整个建筑的重量减少了10%，进而也减少了材料的使用，使得塔楼的结构更透明，与其优雅的形态达成了视觉与概念上的和谐。

Project name: Shanghai World Financial Center
Award date: 2010
Location: shanghai, china
Area: 382,000 m²
Architect: KPF Associates PC, East China Architectural Design & Research Institute, Irie Miyake Architects
Client: Mori Building Company
Photographer: Mori Building Co. Ltd., Michael Moran/Interiors by Tony Chi, Tim Griffith, H.G. Esch
Completion date: 2008
Award name: McGraw-Hill Construction 3rd Bi-Annual "Good Design Is Good Business" China Awards 2010, Best Commercial Project

项目名称：上海环球金融中心
获奖时间：2010
项目位置：中国 上海
项目面积：382,000平方米
建筑设计：KPF建筑师设计事务所　华东建筑设计研究院　入江三宅建筑师事务所
业主：日本森大厦株式会社
摄影师：Mori Building Co. Ltd.　Michael Moran/ Interiors by Tony Chi　Tim Griffith　H.G. Esch
完成时间：2008
所获奖项：麦格劳–希尔公司《建筑实录》、《商业周刊》第三届"好设计创造好效益"中国奖项 2010最佳商业建筑

1. Lujiazui Financial District
   陆家嘴金融区

2. Entrance of Malls
   商场入口

3. Up View
   仰视图

1. Entrance of Hotel
   酒店入口

2. Observation Deck
   观景台

3. Office Lobby
   办公大堂

4. First Floor Plan
   一层平面图

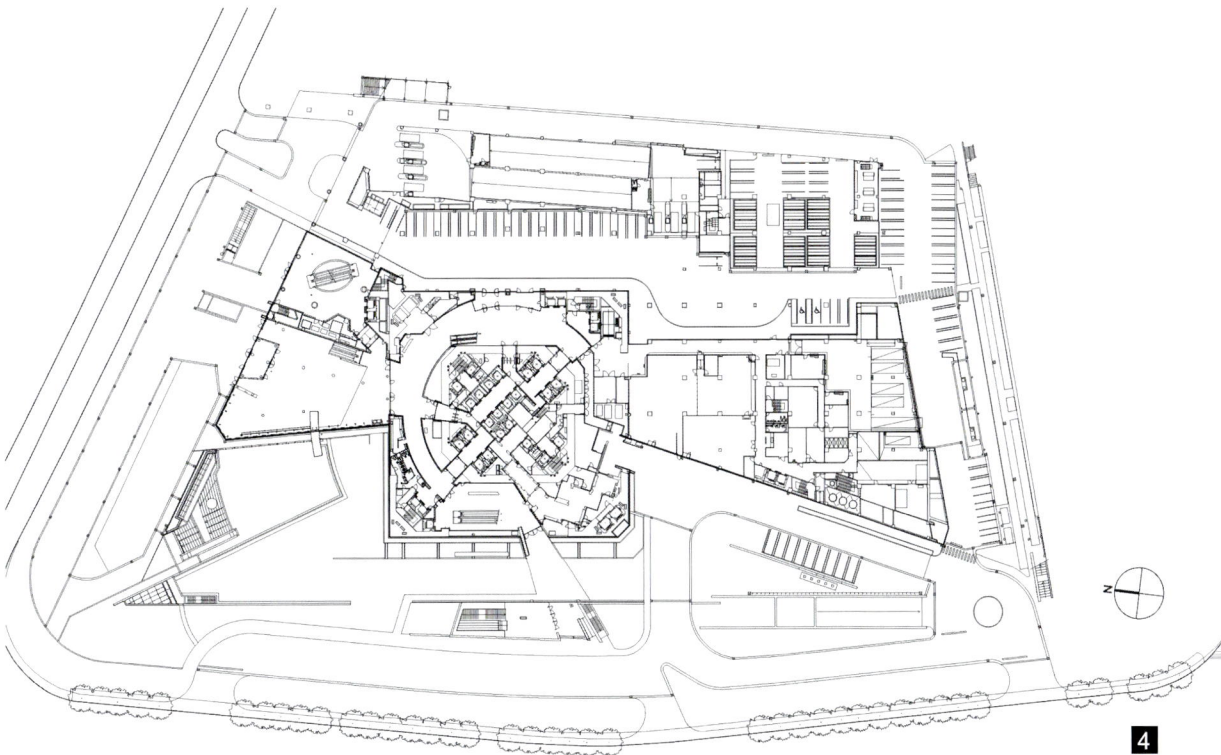

1. Hotel SPA
   酒店SPA

2. Hotel Guest Room
   酒店客房

3. Hotel Bar
   酒店酒吧

4. Hotel Restaurant
   酒店餐厅

Plot 6 of Jishan Base, Jiangsu Software Park

江苏软件园吉山基地6号地块

The Jishan Base, Jiangsu Software Park is located at a piece of beautiful scenery mountains of Jiangning District, Nanjing, consisting of 38 commercial campuses which have several separate office buildings with matching facilities, 1000-2000 square meters. The purpose of the development is to provide the IT enterprises with some suburban work places different from the urban ones.

The single office building uses the courtyard-style layout. A high courtyard wall encircles in the natural fluctuative base, a smooth artificial world where the numerous courtyards and the open-style office space are in perfect harmony, spreading gently on the first story, showing the superiotity of the suburban offices. The second and third stories offices are further arranged to respectively rely on the peripheral courtyard wall, enjoying the suburban genial sunlight, the fresh air and the beautiful scenery fully. Simultaneously this design will not bring the negative influence to a courtyard pleasant criterion. In fact, the kinds of two different density's spaces of the first story and the upper stories achieved a balance of mutual supplement.

The single buildings in the plot are compliant with the topography and the whole organic settlement has the similar structure as the inner space layout in the single buildings. The courtyard wall which construct the internal world also become the boundary between the exterior spaces.

The courtyard walls have white coating while the exterior of the other stories are covered by wooden shading materials under the light sky of Nanjing. The pure white courtyard wall and the calm nigger-brown trellis express the deep respect for the traditional common people residence of the south of Yangzi River.

江苏软件园吉山基地位于南京市江宁区一片风光秀丽的丘陵中，是由38幢面积为1000及2000平方米的独幢办公楼及相关配套设施组成的商务园区，开发目的是为IT企业提供有别于都市写字楼的郊外办公场所。

办公楼单体建筑采用院落式布局。一层高的院墙在自然起伏的基地中围合出一个平整的人工世界，在这里，众多院落与开放式办公空间水乳交融，于一层缓缓展开，诉说着郊外办公的优越之处。二三层办公空间被进一步分解，各自倚在周边的院墙上，充分享受着郊外和煦的阳光、清新的空气和秀丽的景色，同时也不会对一层院落的宜人尺度带来负面影响。事实上，一层和二层以上两种不同密度的空间恰好达成了一种相互补充的平衡。

地块内的各单体建筑结合地形起伏布局，形成一个有机的聚落，这样的布置方式与单体建筑内部的空间组成方式是同构的。院墙在限定建筑内部世界的同时又成为建筑之间外部空间的边界。

建筑的院墙采用白色涂料，而二三层体量的其他几个立面则以木质遮阳构件包裹，映衬在郊外的青山绿水中。在南京淡淡的天空下面，纯净的白色院墙与沉着的深棕色木质格构向江南传统民居表达着深深的敬意。

Project name: Plot 6 of Jishan Base, Jiangsu Software Park
Award date: 2010
Location: Nanjing, China
Building area: 23,000 m²
Architect: Deshaus Architects Company
Client: Jiangsu Xingyuan Real Estate Development Co., Ltd.
Photographer: Shu He
Completion date: 2008
Award name: McGraw-Hill Construction 3rd Bi-Annual "Good Design Is Good Business" China Awards 2010, Best Commercial Project

项目名称：江苏软件园吉山基地6号地块
获奖时间：2010
项目位置：中国 南京
建筑面积：23,000平方米
建筑设计：大舍建筑设计事务所
业主：江苏兴园置业发展有限公司
摄影师：舒赫
完成时间：2008
所获奖项：麦格劳－希尔公司《建筑实录》、《商业周刊》第三届"好设计创造好效益"中国奖项 2010最佳商业建筑

1

2

1. Exterior of the Building
   建筑外立面

2. Exterior Space of Roof
   屋顶外部空间

3. Inner Courtyard
   内庭院

4. Plan
   软件园平面图

1

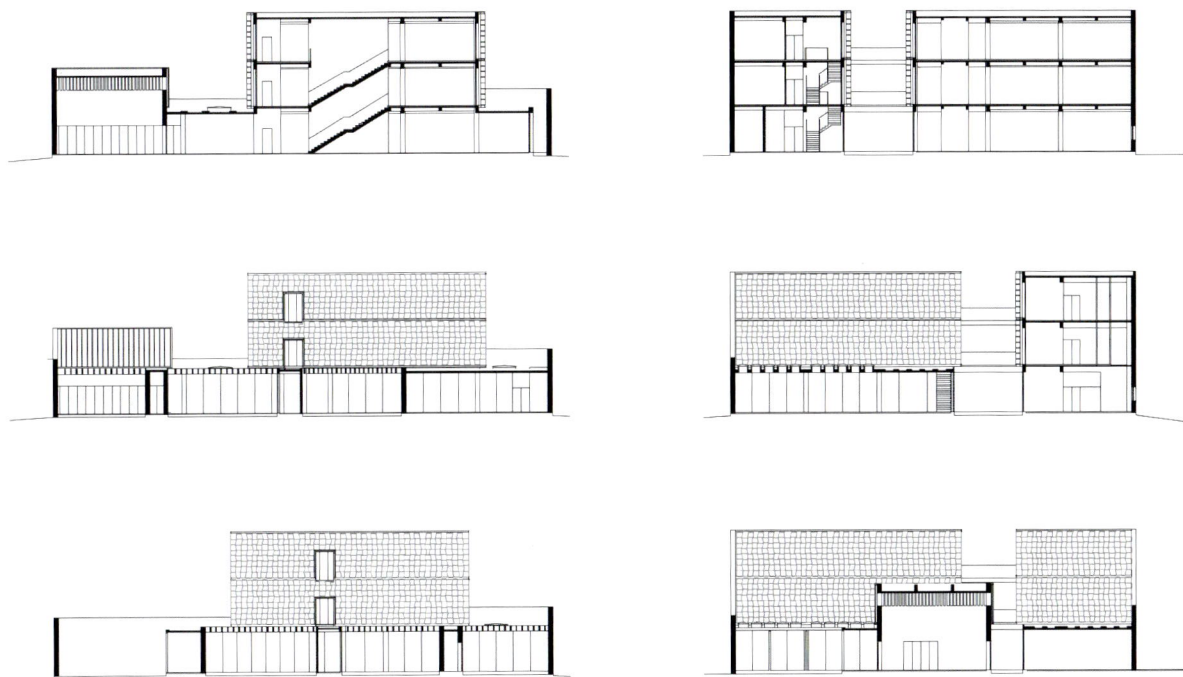

3

1. Full View from the Street
   沿街外观

2. View from Southeast
   东南立面

3. Sections
   剖面图

The Black Box

黑盒子——如恩设计研究室与设计共和办公楼

The concept of the "Black Box" is the guiding concept behind the architecture - modeled after the "black box" flight data recorder, it is used symbolically to represent the "storage" of conversation, ideas, thinking and research in the creative studio office. The black box also serves the function of protecting that recording in the event of a crash, fire or tragedy, analogous to the role of a design office servicing as a container of its intellectual production and protection from outside damage. The black box offers poignant, relevant and passionate design ideas with meaning and purpose to clients who may have had to face design tragedies in their lives. The ground floor in the form of a retail store displays some of these designed objects produced in the offices above, rendering it a window into the contents of the black box.

The Black Box is a five-story office building located in the former French Concession, which also includes a street-level storefront space. On the ground level, two wooden facades make up the base of the building, one comprising the new Design Republic store and the other leading up to the Design Republic and NHDRO offices. The gallery and store on the ground level then become an extension of the street. Above this glass and wooden exterior, a four-story dark facade is extruded and "cut" to reveal windows into the building.

Within the Design Republic space, the wooden box is pierced to reveal white boxes that frame the main display area. Private offices are contained within glass walls, just like within the original Design Republic office on the Bund. The upper two stories will comprise the NHDRO space, which is connected vertically with openings and horizontally with a bridge. The conference room consists of two stacked boxes, a wooden box atop a white box. The room is visible from the upper level through an opening alongside the bridge.

"黑盒子"的概念贯穿整个设计，灵感来自于飞机上记录数据的黑匣子。新办公空间的上面四层楼层代表着对话、想法、思考和研究，与飞行机组人员在飞机上的谈话极其相似。 当飞机发生撞击、火灾或类似悲剧的时候，黑匣子记录了所有数据。黑盒子用同样深刻、贴切并充满激情的设计理念，寓意了我们也将帮助那些需要在人生中面对设计悲剧的客户们。位于一楼的零售店则展现了那些随着时间推移被储存起来的想法。

余庆路88号原本是法租界内的一栋五层办公楼。底楼入口两侧的木制外立面构成了大楼的基础，其中一侧延街道舒展，与玻璃隔断组成设计共和新展厅的外墙，另一侧则向内延伸，指引着通向楼上办公空间的电梯与楼梯入口。底层敞开式的展厅成为街道的延伸，二楼之上的深色外立面仿佛是突出之后的裁切，密密麻麻的四方形窗口极具立体感的显露出来。

在新的设计共和空间内，木制"盒子"被穿透，显露出构成展示区域的白色"盒子"。在二楼，纯白的陈列室里有一个厨房，里面装配整套Miele的设备。进入三楼设计共和的办公空间要穿过一个橡木"盒子"。个人办公室采用玻璃墙隔断，和原在外滩5号的设计共和办公室一样。

四五两层是如恩设计研究室的办公空间，垂直方向上打通楼面贯穿，水平方向上则是一道桥梁连接。在五楼的"桥边"透过玻璃墙俯视，便是四楼的会议室 —— 那是由叠起来的两个"盒子"组成的房间，木盒子在上，白盒子在下。

Project name: The Black Box
Award date: 2010
Location: Shanghai, China
Building area: 1,500 m²
Architect: Neri & Hu Design and Research Office
Client: Neri & Hu Holdings Limited
Photographer: NHDRO, Derryck Menere, Tuomas Uusheimo
Completion date: 2009
Award name: McGraw-Hill Construction 3rd Bi-Annual "Good Design Is Good Business" China Awards2010, Best Commercial Project

项目名称：黑盒子——如恩设计研究室与设计共和办公楼
获奖时间：2010
项目位置：中国 上海
建筑面积：1,500平方米
建筑设计：如恩设计研究室
业主：如恩控股
摄影师：如恩设计研究室，Derryck Menere, Tuomas Uusheimo
完成时间：2009
所获奖项：麦格劳-希尔公司《建筑实录》、《商业周刊》第三届"好设计创造好效益"中国奖项 2010最佳商业建筑

1. Grand Stair
   主楼梯
2. Elevator
   电梯
3. Floor Reception
   前台
4. Kitchen
   厨房
5. Server Room
   服务室
6. Main Meeting Room
   主会议室
7. Model Display
   样品陈列
8. Restroom
   休息室
9. Print Room
   打印室
10. Graphics Department
    版式部
11. Pin-up Wall
    可贴纸的墙面
12. Open Studios
    开放式工作区
13. Critique Nook/ Book Library
    图书阅览室
14. Materials Library
    材料室

1. View from the Sidewalk
   沿街外观
2. Plan
   平面图
3. Entrance
   入口
4. Design Republic New Showroom
   设计共和新展厅
5. View from the Sidewalk
   沿街外观

1

2

3

1. Fifth Floor Entrance
五层入口

2. Ground Floor
一层

3. Fourth Floor Meeting Room
四层会议室

4. Third Floor Conference Room
三层接待室

5. Fifth Floor Entrance
五层入口

1

2

1. Corridor
   走廊

2. View from the Bridge
   桥上观景

3. Fourth Floor Meeting
   Room
   四层会议室

3

Jianianhua Center

嘉年华中心

Chongqing is a foggy riverfront city that sought to breathe new life into its commercial district, Jiangbei. As part of a redevelopment master plan, the city commissioned the design of a central park and plaza, which would be anchored by the Jianianhua building. The client initially sought a super-tall iconic tower, but through discussions about their goals and studies of the district, the design team became convinced that a lower building would be more in scale with the park, less imposing, yet equally powerful as a civic landmark. In addition, cars were banished to the perimeter, redefining the area as a pedestrian haven.

The Jianianhua building itself defines a new kind of civic landmark, one that establishes meaning and presence through an innovative integration of architecture and graphic design. The transparent glass building (40,000 square meters of retail and 10,000 square meters of office space, plus parking) steps down toward the park, with seven floors of retail organized in a C-shape around an atrium and a slender office tower raising another seven stories behind. At the same time, the building appropriates the large-scale commercial signage that predominates in the city and elevates this mode of urban communication to the level of public art.

As the building neared completion, the client commissioned SOM to design its first graphic, which was unveiled as a gift to the city at the Chinese New Year festival in 2005. Buildings have always communicated the values and aspirations of their time, and the Jianianhua Center embraces this moment for Chongqing. It blurs the line between building and media and combines the possibilities of our information-powered culture with advanced building technology, creating an evolving architecture that defines a significant public space and connects a city with its people in a radically new way.

重庆多雾，素有"雾都"之称，目前着力于打造江北商业圈，为重庆市城市面貌的改善增添一道魅力风景。中央公园和广场作为规划的一部分，建成后与嘉年华大厦相映成趣。客户起初拟建一个超高层标志性建筑，在对项目所在地的认真考察和仔细研究之后，决定将楼层降低，以实现与公园更完美地契合。同时，这里禁止机动车的出入，是一个理想的步行天堂。

嘉年华大厦轻盈飘逸的建筑外观、通透现代的建筑风格和追求卓越的建筑品质使其成为该地区的标志性建筑，吸引了无数观光者驻足欣赏。建筑的外观采用透明玻璃结构（设有40,000平方米的零售店，10,000平方米的办公空间以及停车场），位于中心的"C"形7层商业大楼与公园相通，其后侧是一个狭长的商务办公大楼。整个空间以商业办公功能为主，同时为公众提供极佳的休闲娱乐空间。

应客户要求，设计师将在2005年初提供首个形象设计，作为农历新年的节日礼物呈现给市民。

建筑往往能够彰显出同时代的价值观和文化特色，嘉年华大厦亦是如此。它淡化了建筑与媒体之间的界限，将信息文化与先进的建筑科技文化融为一体，打造城市建筑新景观的同时，拉近了城市与公众的距离。

Project name: Jianianhua Center
Award date: 2006
Location: Chongqing, China
Project area: 64,340 m²
Architect: Skidmore, Owings & Merrill LLP
Client: Chongqing Financial Street Real Estate Co.
Photographer: Courtesy of Skidmore, Owings & Merrill LLP, Tim Griffith
Completion date: 2005
Award name: McGraw-Hill Construction 1st Bi-Annual "Good Design Is Good Business" China Awards 2006, Best Commercial Project

项目名称：嘉年华中心
获奖时间：2006
项目位置：中国 重庆
项目面积：64,340平方米
建筑设计：SOM建筑设计公司
业主：金融街重庆有限公司
摄影：SOM建筑设计公司 Tim Griffith
完成时间：2005
所获奖项：麦格劳-希尔公司《建筑实录》、《商业周刊》第一届"好设计创造好效益"中国奖项 2006最佳商业建筑

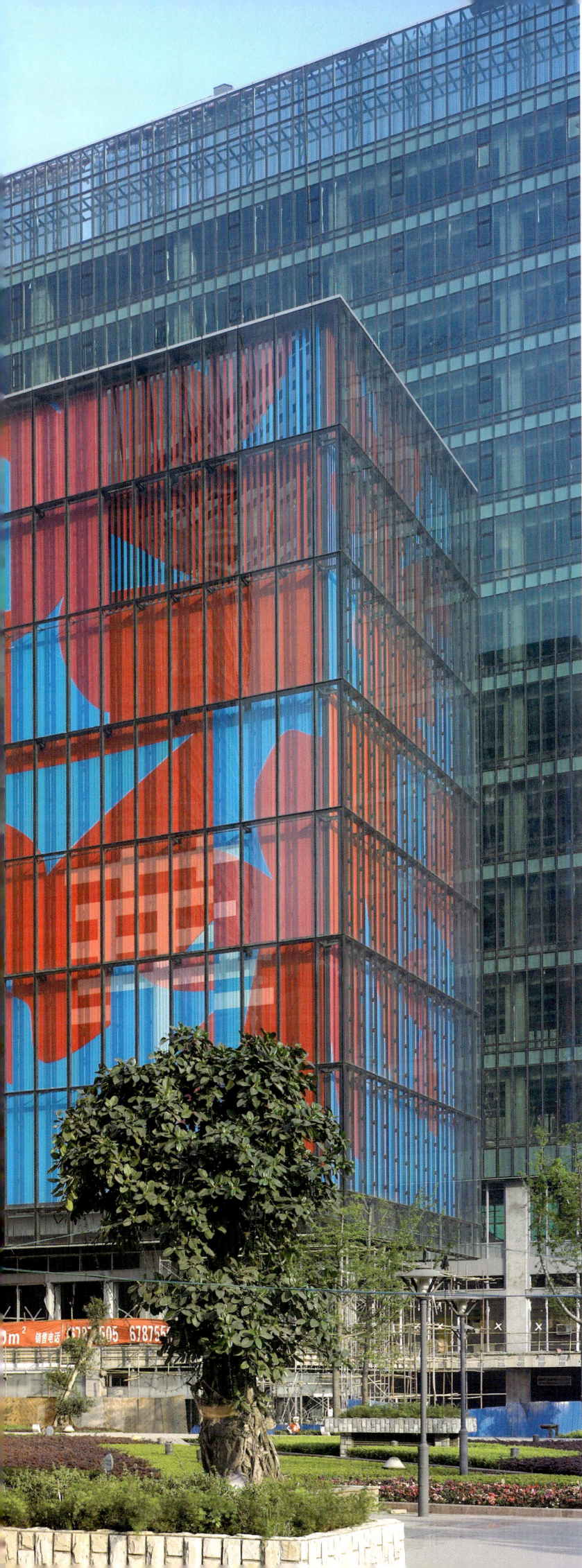

1. Night View
   夜景

2. Perspectives
   透视图

3. Night View
   夜景

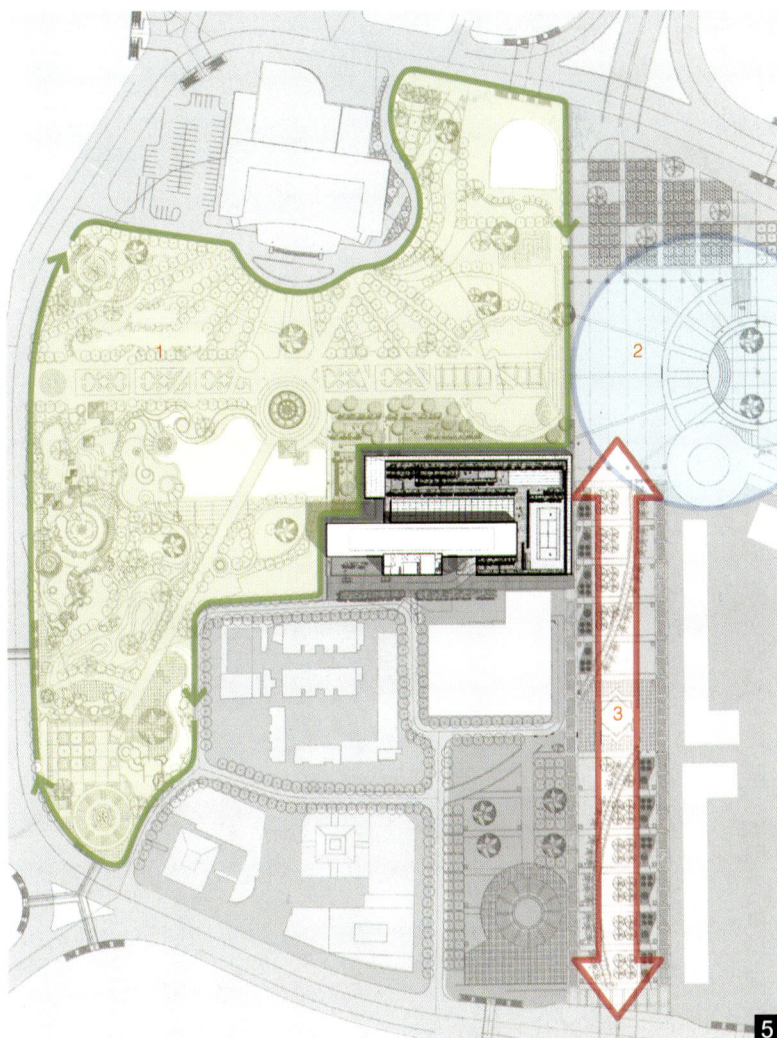

1. Facade
   正面图

2. View from Far away
   远景

3. Detail Part
   细部图

4. Interior
   室内

5. Plan
   平面图

1. Park Promenade
   公园

2. Civic Plaza
   城市广场

3. Pedestrian Shopping Street
   步行商业街

Qingpu Private Enterprise Association Building

青浦私营企业协会办公楼

Located at the east side of Xiayang Lake in the new area of Qingpu District, the office building belongs to the Xiayang Lake Landscape Area.

The designer took into consideration the views not only from the interior but also from the exterior. The solution is a cube covered with a glass curtain wall. The cube building is 60x60 meter. The three-story glass wall encloses a green yard that creates a boundary offset the building at least 4 meters. While obviously defining the private space in the interior, the transparent glass walls also create a visual communication to the landscapes inside and outside. The main buildings are arranged into a square shape with a courtyard at the center. The first floor is elevated and only houses the reception and a restaurant. In this way, the ground floor spaces flow from the central garden to the exterior landscape and an uninterrupted view goes through the building.

The glass walls work as a boundary, and create a microclimate in this area. Firstly, they largely reduced the noise from the highway from east. In Chinese Garden, the ponds and trees in the courtyard are the significant issues to create a microclimate. During the summer, the vapor from the pond with the elevated floor provides an air circulation that reduces the temperature in the building. When visitor goes into the building, a sudden cooling from the garden brings a feeling of serenity.

Glass panels are hanging on the facade with stainless steel fixings, fixing glass on the right, left and bottom. Each panel has a gap in between. We use screen printing for the curtain glass wall of the inner main structure. They create wholeness visually, and also work as a shading system. The pattern of screen printing that we choose is a shape of "broken ice" and the image of dragonfly's wings.

In the interior design, white color is used as a dominant color, white artificial stone, white paint on the metal trussed roof, glass and timber. Those modern materials with its significant atmospheres bring imaginations.

这座办公楼位于青浦新城区中心夏阳湖的东侧，在总体区位中处于特定位置的这座办公楼可以为整个新城区规划结构的完整性作出贡献。

一个60米见方的三层通高的玻璃围墙为这座建筑建立了一个明确的边界，这层"边界"和内部的建筑外墙空出了至少4米的距离，玻璃围墙的通透性同时也使得这个"边界"在建立后自我消解。玻璃墙内部的建筑主体围绕一个更为内层的庭院大致呈"口"字形布局，同时底层大部分予以架空，二层朝向主要景观（如西北方向的湖面和东面的小河）的部分设置了高敞的观景平台，这使得玻璃围墙内庭院的绿化与外部环境的绿化、景观形成了相互渗透。

玻璃围墙也是一个调节或过滤了的微气候提供了一个恰当的边界。它首先大大减弱了来自东部不远处高架高速公路的噪音干扰，而在江南地域，庭院中的水池和树木对庭院的小气候也起着不可估量的作用。在炎热的夏季，中心内庭的水池因为水分蒸发消耗了庭院内的部分热能并在周边的架空层下形成微妙的空气流动，当人们从建筑的外部进入围墙内部的架空层下，突然的荫凉带来内心的平静，一切刻意或无意的设计都显得不再重要。

建筑物的外围墙玻璃被一片片悬挂在其上方的不锈钢方形锁板上，每片的左右及下方锁板起到玻璃定位作用。玻璃块与块之间留出缝隙。内部主体部分整体采用了丝网印刷的玻璃幕墙，使建筑体量在视觉上有完整感，并对内部起到一定的遮阳作用。印刷图案选用冰裂纹与蜻蜓翅状网纹，两次叠加印刷。室内设计采用了白色的基调，材料运用了白色的长条人造石，白色喷漆金属隔栅的吊顶，玻璃与木材。走入建筑或庭院内部，这些现代材料和具有某种特质的氛围，使人们在物体本身和整体指向之间游移。

Project name: Qingpu Private Enterprise Association Building
Award date: 2006
Location: Shanghai, China
Building area: 6,745 m²
Architect: Alieter Deshaus
Client: Qingpu Branch of Shanghai Industry and Commerce Administration Bureau
Photographer: Zhang Siye, Hu Wenjie
Completion date: 2005
Award name: McGraw-Hill Construction 1st Bi-Annual "Good Design Is Good Business" China Awards 2006 , Best Commercial Project

项目名称：青浦私营企业协会办公楼
获奖时间：2006
项目位置：中国 上海
建筑面积：6,745平方米
建筑设计：大舍建筑设计事务所
业主：上海工商行政管理局青浦分局
摄影师：张嗣烨 胡文杰
完成时间：2005
所获奖项：麦格劳—希尔公司《建筑实录》、《商业周刊》第一届"好设计创造好效益"中国奖项　2006最佳商业建筑

1

2

1. Hollow Space on the First Floor
   首层架空

2. View from West
   西侧外观

3. Entrance Terrace on the Second Floor
   二层入口平台

4. Entrance Terrace on the Second Floor
   二层入口平台

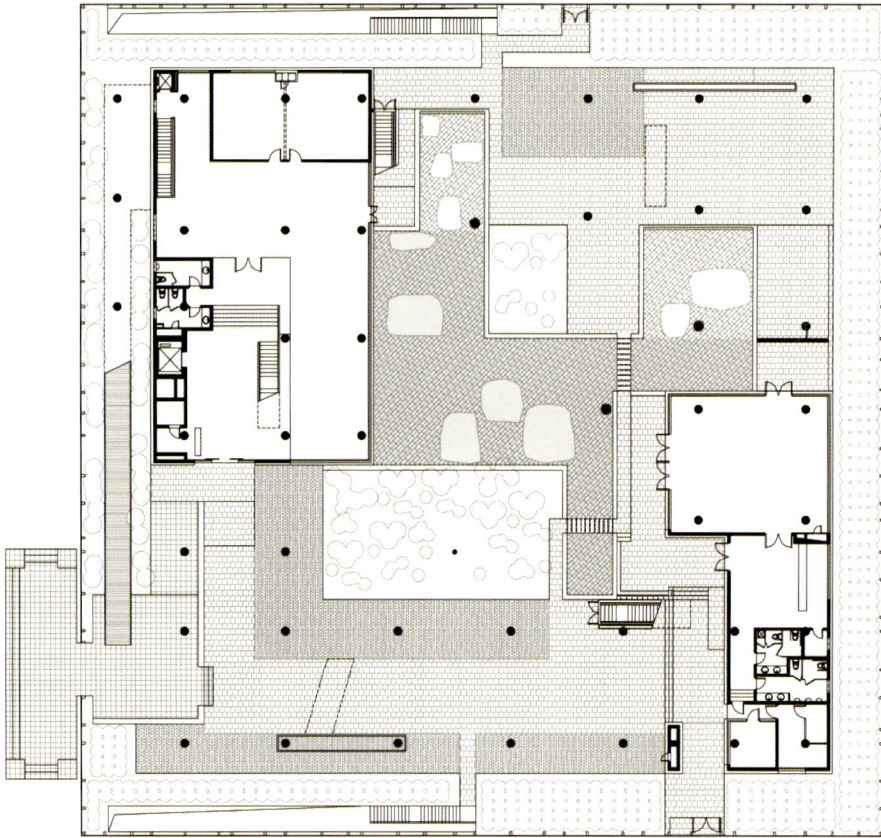

1

2

1. Plan
   平面图

2. The Bridge of the Entrance
   入口坡道

3. Interior Hall
   室内大厅

Tangshan City Hall

唐山市城市展览馆

If the great 1976 earthquake in Tangshan is an unintentional destruction in its architectural history, then the Urbanization Movement deliberately denies to the contemporary architecture of Tangshan. The original flourmill in Tangshan is the victim. When the plant was relocated, its dozens of homely warehouse were planed to be completely turned down and turned into the City Park. For a city where majority of the buildings is only 30 years old, the four warehouses built during the Japanese invasion should be reserved and created into a core of museum group that is built at the foot of the mountain. The six parallel buildings are perpendicular to the mountain and lead the mountain to the city in a rhythm. The new construction gives more prominent of this form, and tries to make the original building and mountain to be the focus. The new construction with rich functional areas, such as the reception room, bookstore, cafe, gift shops, makes much space for the mountain with small volumes and parallels to the original building. Materials only refer to the transparent metal grill and preservative wood in order to strengthen the spirit and natural feeling of the site. Against the alternative but also simple materials, the intact walls of the old warehouse reveal its inner beauty. Along the park side, each warehouse is added a steel porch so as to inject the closed warehouse with a sense of openness. The reflection of the porch in the pool has highlighted the beauty of the old building. The "X"-shaped steel structure roof has transformed the closed room into a bright, perfect, and standard showroom. Currently, this museum complex is defined as the city exhibition hall which tells the story about the city. What's more, in the park, Taobao village, games circle and other public facilities are also created to encourage the public to participate more easily. Landscape design is to use the natural beauty to decorate the building, to create a relaxing cultural environment, thus creating a sense of intimacy. In such a city, this park will provide the public a good place to read more about the history of Tangshan.

如果说1976年的大地震对唐山的建筑史是一次无意的摧毁，那么，今天的城市化运动则是对唐山平庸的现当代建筑史的一次有意的抹杀。原唐山面粉厂便是这样的牺牲品。当工厂外迁后，它的数十座相貌平平的仓库，由于几乎毫无美学价值，计划中被彻底推掉，变成城市公园。对于一个大多数建筑只有30余年的城市，厂区中四幢日伪时期建的旧库房似乎很值得保留。在多方努力下，它们和另两幢80年代建的粮仓得以保存，并以此为核心，形成一个山脚下的博物馆群。这六栋平行的建筑恰巧垂直于山体，它们有节奏地将山引到城市。新加建的部分更加突出了这种天作之合，所以内容非常有限，尽可能使原建筑和山体成为视觉的主体。用来丰富功能活力的新建筑，例如接待室、书店、咖啡、礼品店等，尽可能用小体量来让出山体，并平行于原建筑。材料使用上也很节制，只用通透的金属格栅和防腐木板，来强化场所固有的工业化精神和自然的面貌。在这些另类些、却也朴素的材料映衬下，原封保留的旧仓库的墙面透出了内在的美。沿公园一面，每个仓库增添出一个钢结构门廊，让封闭的仓库具有一种开放性。这些门廊落在反射水池上，使旧建筑的美进一步放大。"人"字形仓库的屋面用"X"形钢结构来代替，形成的侧高窗使原先封闭的室内变成明亮的、非常完美和标准的展示厅。目前这个博物馆群被定义为城市展览馆，向市民讲述城市的故事。从有生机的公园活动的角度，公园内还规划了淘宝村、游戏圈等公众更易于参与的内容。景观设计是用自然美来装点朴素的建筑，以形成一种轻松的人文环境，从而创造一个与一般市民没有距离感的公共空间。在一个似乎一切都很平淡的城市，人们在这个公园里可以不经意地读到唐山沉淀的历史和值得关心的历史残片。从这些历史中，找回对自己城市的信心。

Project name: Tangshan City Hall
Award date: 2010
Location: Tangshan, China
Building area: 5,900 m²
Architect: Urbanus Architecture and Design
Client: Tangshan Urban Planning Administration Bureau
Photographer: Yang Chaoying
Completion date: 2008
Award name: McGraw-Hill Construction 3rd Bi-Annual "Good Design Is Good Business" China Awards 2010, Best Public Project

项目名称：唐山市城市展览馆
获奖时间：2010
项目位置：中国 唐山
建筑面积：5,900平方米
建筑设计：都市实践设计有限公司
业主：唐山市规划局
摄影师：杨超英
完成时间：2008
所获奖项：麦格劳-希尔公司《建筑实录》、《商业周刊》第三届"好设计创造好效益"中国奖项 2010最佳公共建筑

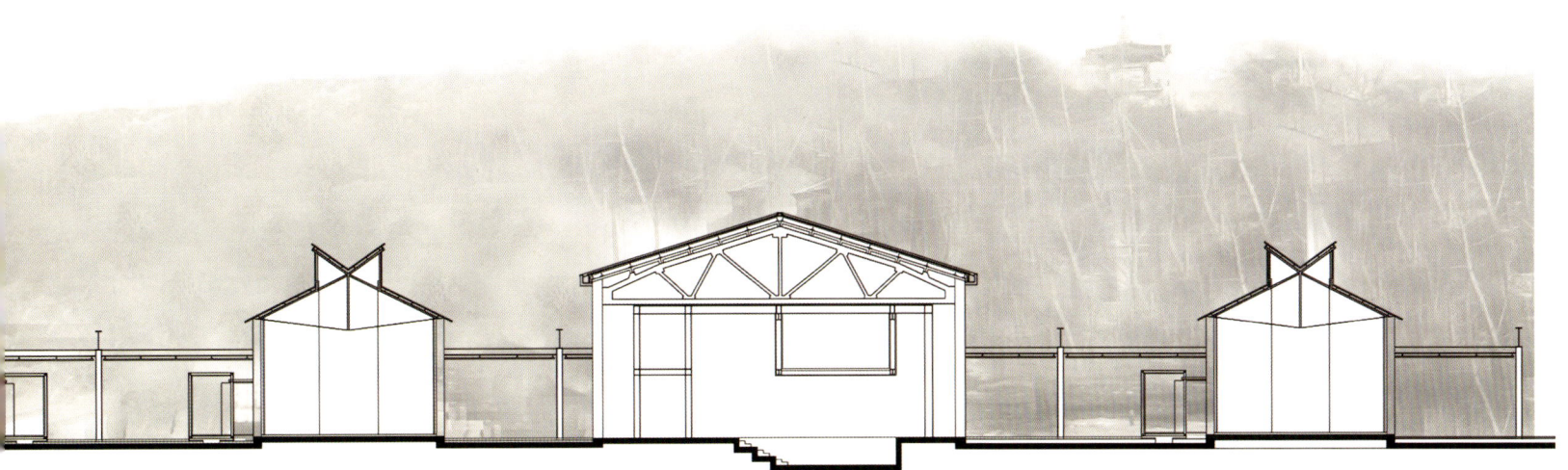

1. Night View
夜景

2. Pool
水池

3. Section
剖面图

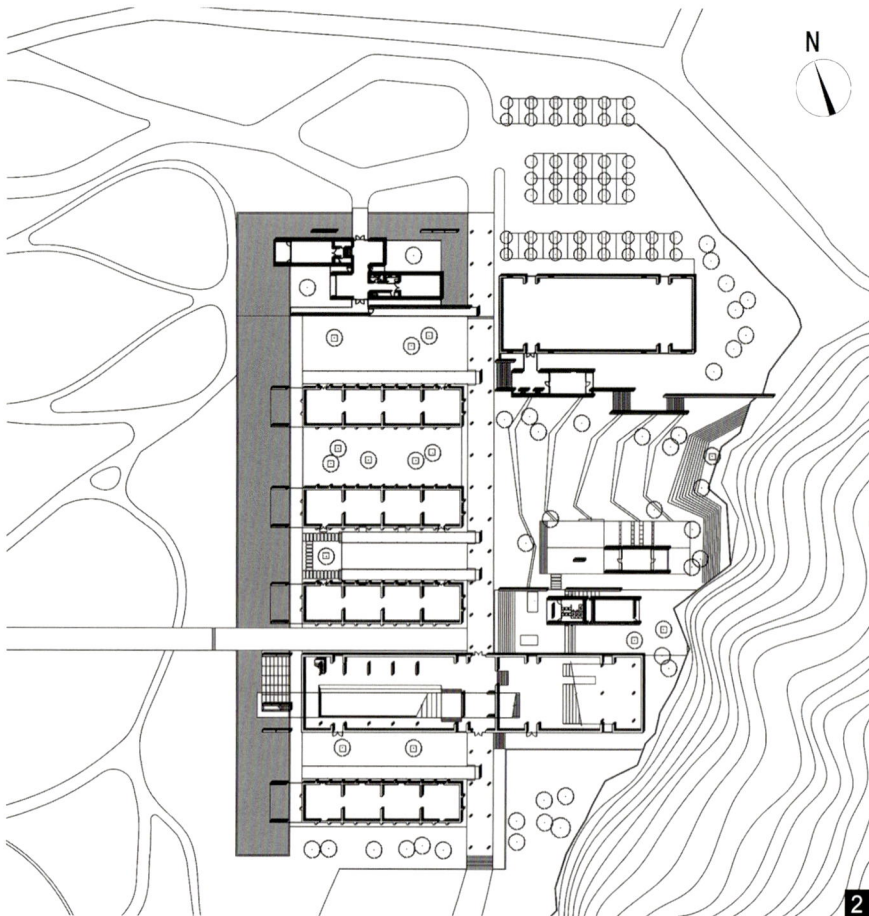

1. Facade of the Front Part
   前廊正视图

2. Plan
   平面图

3. Path Way and Public Area
   步行道及开放空间

4. Exterior
   外部

1

1. Interior
   室内

2. Different Point of View
   侧视图

3. New Material Path
   新材料结构步行道

The OCT Art & Design Gallery

华·美术馆

The OCT Art & Design Gallery was transformed from old warehouse of the Shenzhen Bay Hotel that was built in the early 1980s. The existing building is located in the south of Shennan Road, Shenzhen, adjacent to the famous He Xiangning Art Museum and Shenzhen OCT Intercontinental Hotel. Considering this strategic location, the owner decided to retain and transform it into the hotel's art gallery. The renovated building contrasts sharply with the existing buildings at both sides of the road and pays more attention to the integration with the surrounding environment. A hexagonal glass curtain wall, which is made by overlapping steel structures, forms a shell around the original building and a strong visual impact. The interior space continues to use the hexagonal elements as the facade stretched at 90° in the vertical direction and thus form a series of folded planes across each other and finally constitute complicated but functional public spaces. The renovation has successfully changed the original monotonous geometric pattern, and the use of three-dimensional manner has generated a new and fresh interior space.

建设中的华侨城洲际大酒店展馆原是建于20世纪80年代早期的深圳湾大酒店的洗衣房。在高速发展的城市中，虽一同并列于深南大道南侧，这座存在于华侨城西班牙风情主题酒店和典雅的何香凝美术馆夹缝中的旧厂房，因其单调的建筑形式早已成为不为人留意的都市残留物。厂房的所属方考虑到其优越的地理位置，决定将其保留并改造为附属于酒店的艺术展馆。它虽是邻近的国家级美术馆为展览空间的延伸，但酒店展馆的特殊定位，决定了改造后展馆的独特性：其设计既要突显个性，与两边建筑风格形成差异性对比，同时也要体现与两端建筑的关系及整体性。改造策略完整地保留了原建筑立面的窗墙体系，加建的立面通过包裹的手法，将单一的原始六边形通过复杂有机的组合形成由实至虚，由小到大，多层次渐变的三维视觉效果。从而，在车辆由西至东快速通过的瞬间，形成强烈的视觉冲击力，通过立面结构的缩小放大，逐层递减，如同面纱般轻轻揭开，最终透出原建筑立面的戏剧性变化过程。展馆的室内设计再次运用立面所含有的六边形元素作为基本平面形态，在竖向上作90°的拉伸，形成一系列折叠平面互相交叉、互相切入构成的复杂但带着明确功能元素的公共空间。这种表达形式改变了原本单调的立面几何图案，用三维方式生成新的室内空间，这种"突变"形式使设计产生了不可预料的惊喜结果。

Project name: The OCT Art & Design Gallery
Award date: 2010
Location: Shenzhen, China
Total area: 2,620 m²
Architect: Urbanus Architecture and Design
Client: Shenzhen OCT Real Estate Co., Ltd.
Completion date: 2008
Award name: McGraw-Hill Construction 3rd Bi-Annual "Good Design Is Good Business" China Awards 2010, Best Public Project

项目名称：华·美术馆
获奖时间：2010
项目位置：中国 深圳
总面积：2,620平方米
建筑设计：都市实践设计有限公司
业主：深圳华侨城房地产有限公司
完成时间：2008
所获奖项：麦格劳–希尔公司《建筑实录》、《商业周刊》第三届"好设计创造好效益"中国奖项 2010最佳公共建筑

1. New Exterior of East
   加建东立面
2. Original Building
   原有建筑
3. New Secondary Entry
   加建次入口装置
4. New Exterior of North
   加建北立面
5. New Exterior of South
   加建南立面
6. New Corridor to Hotel
   加建连廊通往酒店
7. New Entry for Administration
   加建后勤入口装置
8. New Exterior of West
   加建西立面
9. New Main Entry
   加建主入口装置

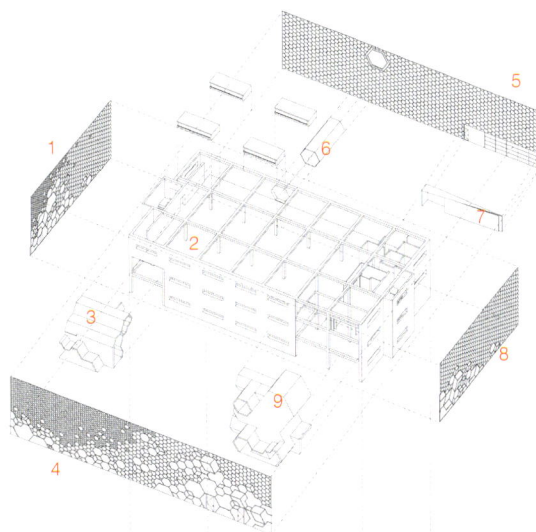

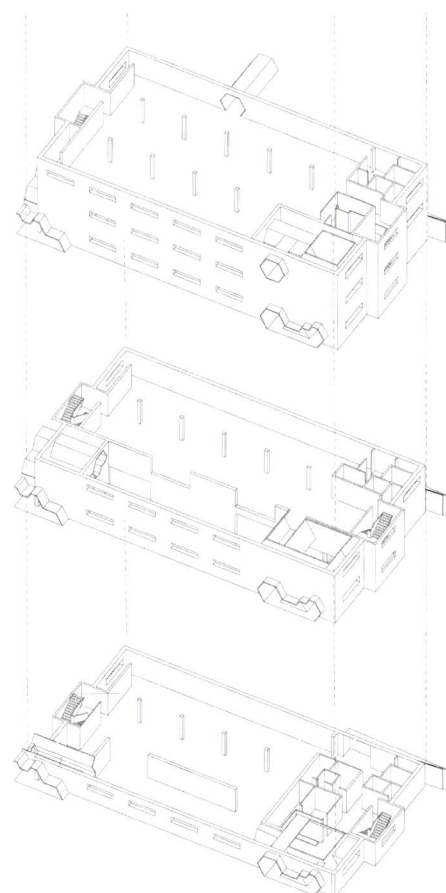

1. Interior
室内

2. Reception Desk
前台

United States Embassy in Beijing
美国驻北京大使馆

At 500,000 square feet, the new U.S. Embassy in Beijing is the second largest Embassy compound ever undertaken by the United States government. Located on a ten-acre site northeast of the Forbidden City in Beijing's new Third Embassy District, the new Embassy is a multi-building campus punctuated by gardens and art. The design emphasizes environmental sustainability and presents an open, gracious and civic face to the city of Beijing. Over 700 Embassy personnel are accommodated within a secure and socially engaging workplace.

The importance of the relationship between China and the United States is reflected by the care taken in the architectural selection process, which emphasized design excellence. The State Department endeavored to select a design that would represent "the best in American architecture" through an invited, national design competition. In making its selection, the design competition jury praised the winning scheme by Skidmore, Owings & Merrill LLP for its "innovative, modern design that respects China's cultural and environmental qualities while honoring and expressing American values through architectural means".

The Embassy is organized into three "neighborhoods": a social neighborhood with retail and recreation spaces, a professional neighborhood with office space, and the Consular neighborhood. The Embassy's Consular Pavilion serves as America's front door to China. Chinese visa applicants cross a lotus garden pond on a stone-framed wooden bridge, arriving at a generous portico – an American front porch.

An ethos of sustainable design underlies the overall Embassy design. The Consular Pavilion directly conveys America's commitment to global sustainability to the Chinese public. The Pavilion uses thermal mass to minimize peak energy demand, resting within one of the lotus ponds which hold and purify storm water. The Pavilion's luminous roof floods its interior spaces with natural light through a set of baffled skylights. The generous use of natural light defines the public and employee interior spaces throughout the Embassy.

The largest of the buildings, at eight stories high, is perhaps the most innovative of the Embassy's structures. It appears as a luminous tapestry during the day, while at night the entire structure becomes a softly glowing lantern.

项目总面积为40,000平方米，是美国国务院历史上第二大海外建设项目。这一综合建筑群占地4公顷，位于紫禁城的东北部。设计强调环境的可持续性，旨在打造一个开放、亲切的空间，为700多名雇员创造安全、舒适的工作环境。

美中双方对本次项目的设计都极为关注。美国国务院在为期一年的遴选过程中力求找到"最好的美国建筑"。建筑师们把自己的作品集提交给一个由享誉全国的建筑大师和美国国务院官员组成的评审小组进行评估。SOM设计事务所最终在竞标中脱颖而出，其提交的提案融入了中国的土壤并象征性地把东西方的传统结合在一起。

新大使馆分为三个"街区"，分别为：公共小区（包括零售和娱乐空间），业务小区，领事/签证小区。所有的小区都通过花园、庭院、木桥和荷塘相连。

新使馆的建设十分重视能源效率和可持续性发展。通过增强隔热性能降低能源消耗。雨水将被保留在庭院中，并通过一系列的荷塘净化得以重复使用。光滑炫目的屋顶配备最新的机械系统将能源消耗降到最低，同时给雇员提供更为高效的工作空间。

新使馆的中心是一个8层的主办公楼。在一天中不同的时间，这个玻璃的外壳会随着光线的变化而变化。在晚上，玻璃发出的光亮使这座楼犹如一个灯笼或灯塔。

Project name: United States Embassy in Beijing
Award date: 2010
Location: Beijing, China
Site area: 10 acres
Architect: Skidmore, Owings & Merrill LLP
Client: US Department of State
Photographer: Courtesy of Skidmore, Owings & Merrill LLP
Completion date: 2008
Award name: McGraw-Hill Construction 3rd Bi-Annual "Good Design Is Good Business" China Awards 2010, Best Public Project

项目名称：美国驻北京大使馆
获奖时间：2010
项目位置：中国 北京
项目面积：40,000平方米
建筑设计：SOM建筑设计公司
业主：美国国务院
摄影师：SOM建筑设计公司
完成时间：2008
所获奖项：麦格劳－希尔公司《建筑实录》、《商业周刊》第三届"好设计创造好效益"中国奖项　2010最佳公共建筑

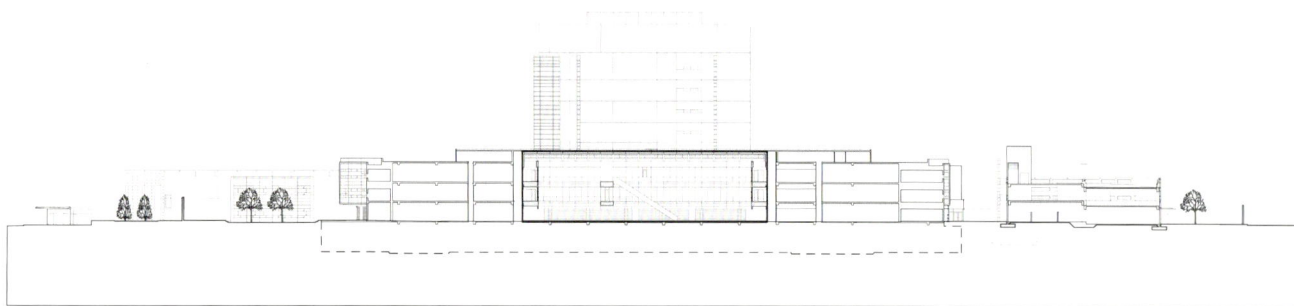

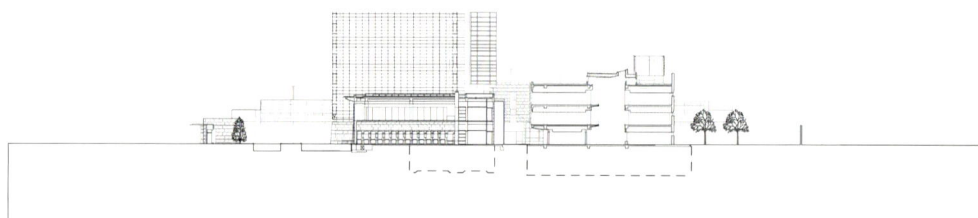

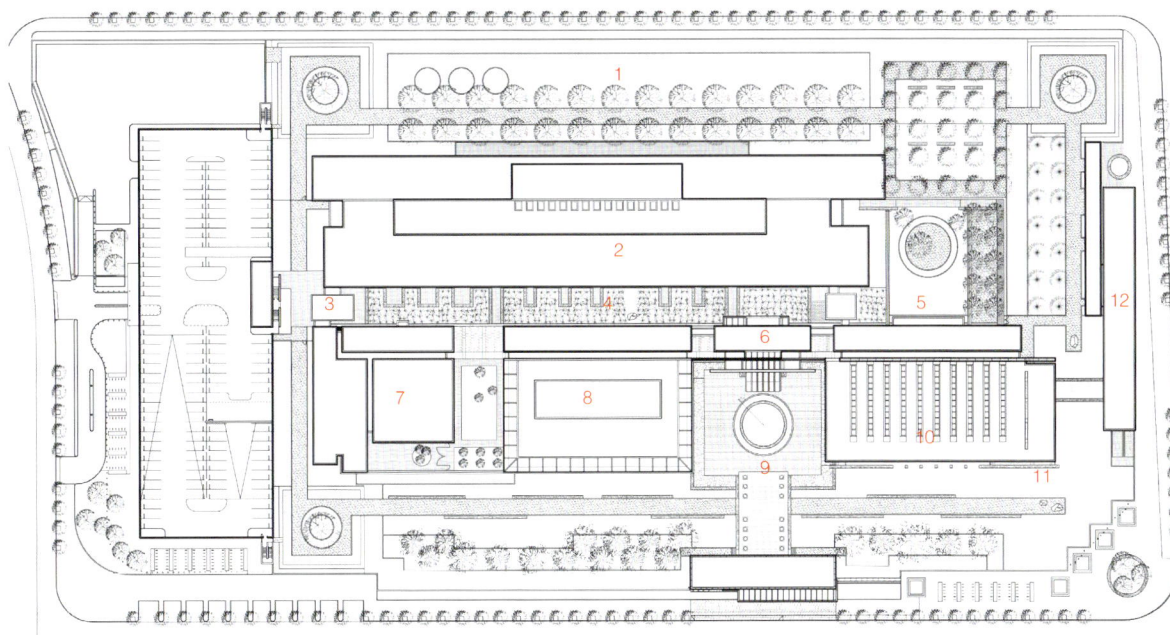

1. North Garden
北花园
2. Administration Building
行政楼
3. Staff Entrance
员工通道
4. Bamboo Garden
竹园
5. Cherry Garden
樱桃园
6. Main Entrance
主入口
7. Recreation and Support
Building
综合楼
8. Chancery
秘书处
9. Entry Court
入口庭院
10. Consular Building
领事楼
11. Parking Structure
停车场
12. Visa Entrance
签证入口

1. Reception Area
   接待室

2. Interior
   室内

3. Interior Detail
   室内细部

4. Escalator
   电梯

Sanghai Qingpu District Exhibition Center of New Town Constrction

上海青浦区新城建设展示中心

Based on topographical features, its minimalistic and extreme linear form is highly consistent with the surrounding large-scale square, lake, TV tower, avenue and playground, and forms an enclosed boundary of Xiayang Lake. From outer zone going inwardly, following a sequence of landscape design of water-body which is close to main road, to a platform for relaxation, to inner garden made of grey bricks, then to inverted architectural space. It expresses a different relationship for opening from urban public space to inner space within architecture, providing a minimalistic volume and rich, delicate details by different treatments of the same of black stone, and transmitting a traditional character by using abstract elements such as color, density, materials and space, etc.

根据地形特点，以简洁超长的线型配合周边的广场、湖面、电视塔、大路、操场等城市性大尺度，形成对夏阳湖的围合性边界。由外而内，以临路水体、休闲平台、青砖院落及凹入建筑的空间等层层递进的景观设计，表达出由城市公共空间到建筑内部的不同开放关系。通过对同一黑色石材的不同加工，赋予简明的体量以丰富细腻的细节。运用色调、密度、材料、空间等抽象元素，传达出江南传统气质。

Project name: Sanghai Qingpu District Exhibition Center of New Town Constrction
Award date: 2010
Location: Shanghai, China
Building area: 10,155 m²
Architect: Jiakun Architects
Photographer: Bi Kejian, Lve Hengzhong
Client: Shanghai Qingpu New Town Construction Company
Completion date: 2006
Award name: McGraw-Hill Construction 3rd Bi-Annual "Good Design Is Good Business" China Awards2010, Best Public Project

项目名称：上海青浦区新城建设展示中心
获奖时间：2010
项目位置：中国上海
建筑面积：10,155平方米
建筑设计：成都家琨建筑设计事务所
业主：上海青浦区新城区建设发展有限公司
摄影师：毕克俭 吕恒中
完成时间：2006
所获奖项：麦格劳-希尔公司《建筑实录》、《商业周刊》第三届"好设计创造好效益"中国奖项 2010最佳公共建筑

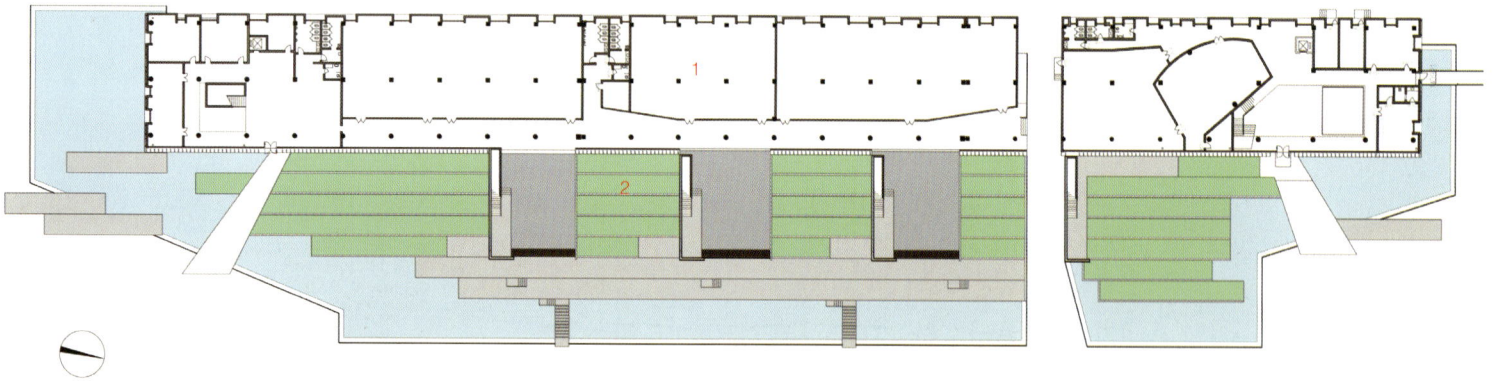

1. Office Building
办公楼

2. Water and Plants
水景和绿地

3

4

1. Full View
   全景

2. First Floor Plan
   一层平面图

3. Entrance and Water Landscape
   入口及水景

4. Exterior Space
   外景

1. Water
   水景

2. Entrance with Green
   入口及绿地

3. Interior Hall
   室内大厅

4. Master Plan
   总平面图

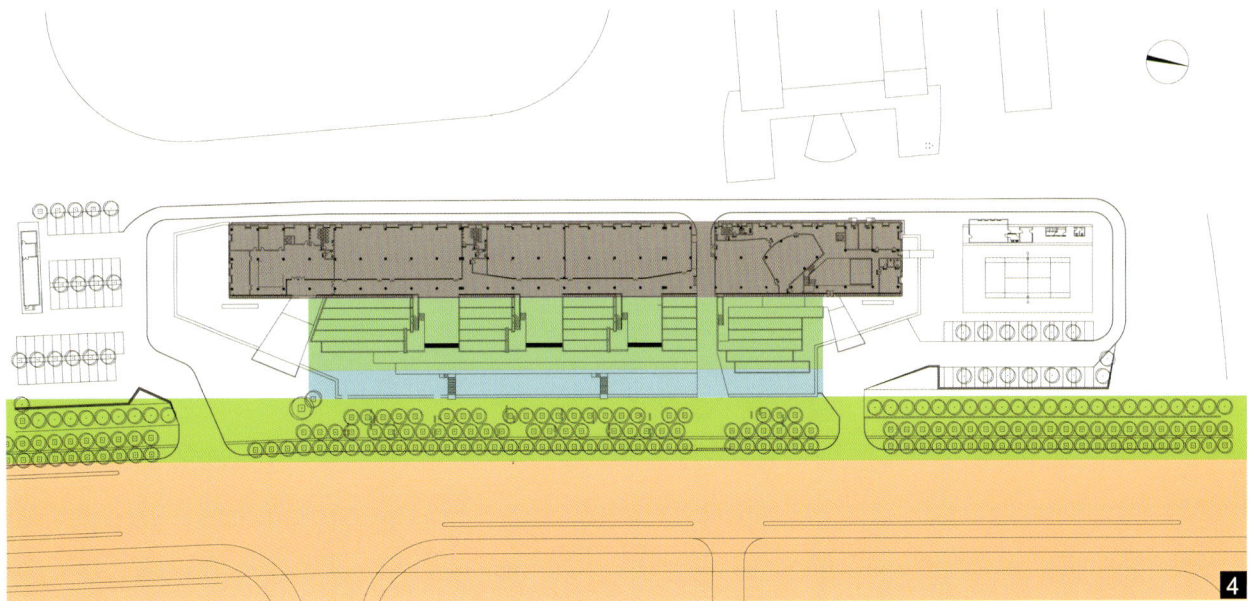

Jinan Sports Center

济南奥林匹克体育中心

Jinan Sports Center is not only an important venue for the eleventh National Games of China, but also an initiating project for the construction of the new town in the eastern part of Jinan City. The Sports Center aims at holding all the sport events of the National Games, and meeting the requirements of General Association of International Sports Federations to provide modern, world-class venues for body building and entertainments. It is a complex that represents the image of the city, and would bring the latter economical benefit.

The plan maximizes the plot ratio and distribution of the venues (a triangle in plan). A sense of balance among spaces, and future business during and after sport events are carefully thought. The stadium and the training venue are situated in the west, while in the east are another stadium, a pool and a tennis center. The western part feels more magnificent, while the eastern layout is much more concentrated, with the central circular stadium and the symmetrical pool and tennis center. The latter two "harbor" the stadium, forming a biaxial symmetry with the stadium in the west. The western, eastern, and southern parts (Administration Center in the south) occupy three corners of the triangle, making a harmonious and steady layout. The central part is the plaza of the National Games, and would become a landmark of Jinan City after the Games. The underground part is devoted to shops, services and parking facilities. The plot, with the south on a higher level than the north, and the east and west higher than the central part, is carefully studied on the site to set an appropriate level for the construction in order to maximally meet the functional requirements of the project. Furthermore, the local culture of Jinan is embedded in the complex.

Jinan Sports Center constitutes a landmark of the district, and even a landmark of the city. Its distinctive forms and perfect functionality would help facilitate the development of the city.

济南奥林匹克体育中心不仅是承办十一届全运会的重要设施，还将成为济南市东部新城建设的启动点。场馆定位为能满足第十一届全运会赛事要求及国际单项运动协会要求的现代化、国内领先的一流比赛场馆，一个提升城市形象，提供康体、健身、休闲一体的、能带回经营回报的生活活动综合体。

规划和设计在用地规模，容积率及"品"字型布局上优化设计，在空间均衡感、满足赛时、赛后的运营设计等进行重点考虑。设计布局在西场区布置体育场及训练场，东场区布置体育馆、游泳馆以及网球中心等。西区的布局流畅、大气；东区的布局紧凑、协调，以圆形体育馆为中心，游泳馆、网球中心以两组对称的体型对体育馆形成环抱，从而与西场区的体育场实现了空间及体量上的双轴对称。同时，西区的体育场，东区的体育馆、游泳馆及网球中心和场地南侧的政务中心形成"三足鼎立"、稳定、和谐的布局形式。中央区域地面为全运会的体育广场，并在赛后成为济南的城市标志性广场，地下为商业、服务及停车等功能。奥体中心用地地势南高北低，东西两侧高，中间低，结合这一体现济南奥体设计特色重要因素，经过仔细现场踏勘验算，场地内的设计标高最大限度地结合了地形，满足使用功能要求，济南奥体中心体育场馆除了满足各项比赛的体育工艺要求外，还充分体现了济南的地域、文化特色。

奥体中心将以其鲜明的建筑造型，完善的功能要求，成为这一地区和济南的重要标志性建筑，为城市的建设和发展带来新的无限生机。

Project name: Jinan Sports Center
Award date: 2010
Location: Jinan, China
Building area: 350,000 m²
Architect: CCDI
Client: Jinan City Development Co., Ltd.
Photographer: Chen Su, Ji Chengke, Li Yan, Zheng Quan
Award name: McGraw-Hill Construction 3rd Bi-Annual "Good Design Is Good Business" China Awards 2010, Best Public Project

项目名称：济南奥林匹克体育中心
获奖时间：2010
项目位置：中国 济南
建筑面积：350,000平方米
建筑设计：中建国际设计顾问有限公司
业主：济南市城市建设投资有限公司
摄影师：陈溯 籍成科 李岩 郑权
完成时间：2008
所获奖项：麦格劳–希尔公司《建筑实录》、《商业周刊》第三届"好设计创造好效益"中国奖项 2010最佳公共建筑

1. Full View
全景

2. Bird View
鸟瞰

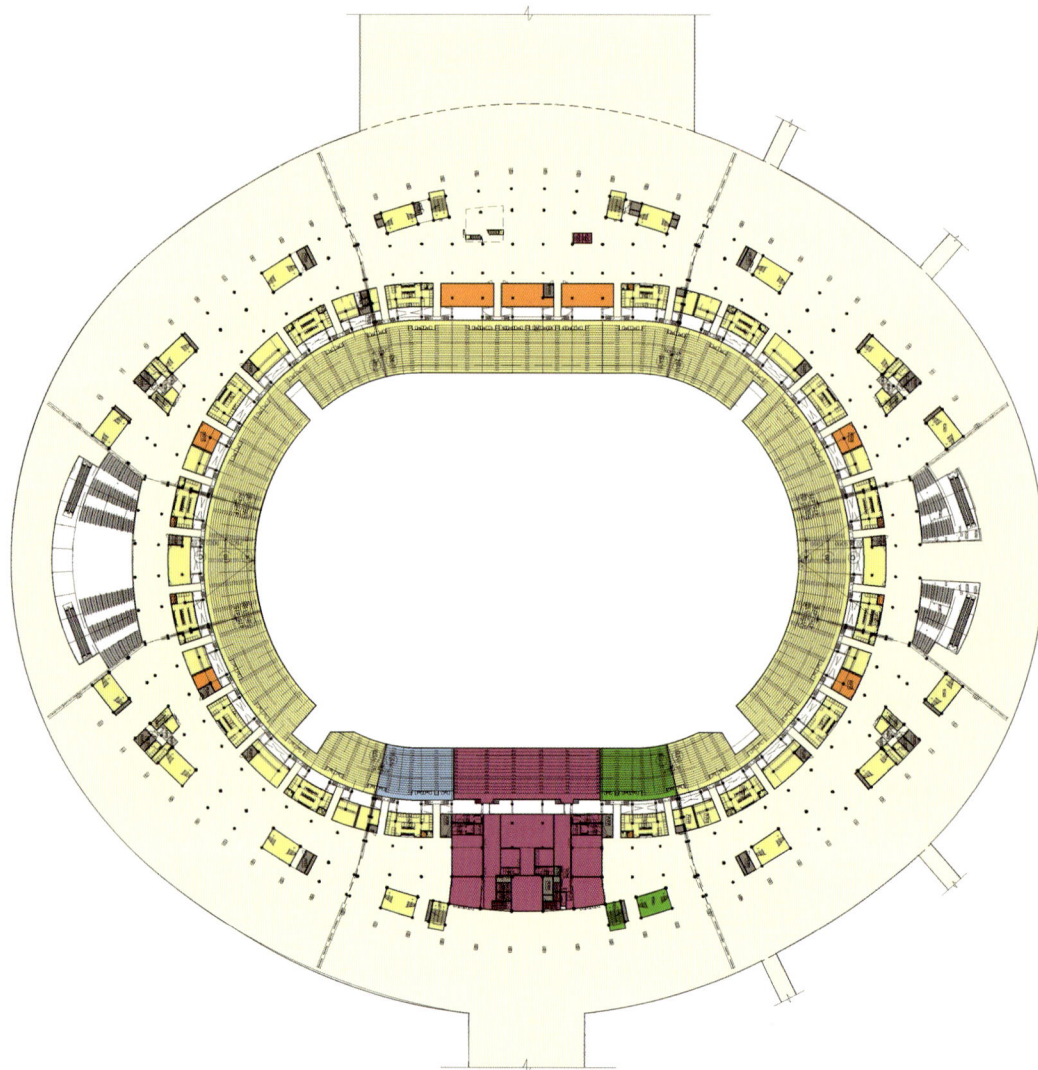

1. Athletes
   运动员
2. Commercial
   商业
3. VIP
   贵宾
4. Press
   媒体
5. Technical Officers
   技术官员
6. Running Department
   场馆运营
7. General Audience
   普通观众
8. Computer Room
   机房
9. Corridors and Transportation
   走廊及交通空间
10. Exterior Terrace
    室外平台

1

2

1. Plan
   平面图

2. Exterior with Water
   亲水外景

3. Detail of the Exterior
   幕墙细部

4. Roof of the Stadium
   体育馆索穹顶

1. Main Court
   田径场地

2. Inner Court
   室内赛场

3. Swimming Pool
   游泳馆

4. Tennis Court
   网球场

1. Hall
   大厅

2. Stairs
   室内楼梯

Liangzhu Museum

良渚博物馆

The museum houses a collection of archaeological findings from the Liangzhu culture, also known as the Jade culture (~ 3000 BC). It becomes the northern focal point of the Liangzhu Cultural Village, a newly created park town near Hangzhou. The building is set on a lake and connected via bridges to the park.

The sculptural quality of the building ensemble reveals itself gradually as the visitor approaches the museum through the park landscape. The museum is composed of four bar-formed volumes made of Iranian travertine stone, equal in 18 m width but differing in height. Each volume contains an interior courtyard. These landscaped spaces serve as a link between the exhibition halls and invite the visitor to relax and rest. Despite the linearity of the exhibition halls, they enable a variety of individualized tour routes through the museum. To the south of the museum is an island with an exhibition area, linked to the main museum building via a bridge. The surrounding landscape, planted with dense woods, allows only a few direct views into the park. The entrance hall can be reached through a courtyard, the centrepiece of which is a reception desk of Ipe wood, lit from above. The material concept consists of solid materials that age well, Ipe wood and travertine stone, and extends to all public areas of the museum.

该博物馆陈列了一系列良渚文化，亦称为玉文化的考古发现，并形成了良渚文化村的北焦点，也是杭州附近新兴的一个公园式城镇。良渚博物馆就坐落在湖边，经由桥梁与公园连接。

博物馆整体的雕刻质量在游客通过园区景观接近博物馆时就已经逐渐显现出来。博物馆由四组伊朗灰石材质的长方形盒子组成，均为18米宽，高度各不相同，并且每一部分都有独立的内部庭院。这些环境优美的空间不仅作为连接各个展厅的通道，同时也提供给游客休息和放松的场所。尽管各展厅呈线形排布，但参观者仍然可以找到多种不同的个性化参观路线。在博物馆的南端是一个与主展馆依靠桥体连接的同样有陈列区的小岛。周围的景观设计，茂密树林，让人只可以对整个园区窥见一斑。

穿过一个庭院即进入博物馆的正门，正中是一个木质的前台，上方燃灯。坚硬的材质包括历史悠久的墙体，木质结构和石灰石，形成了该项目的材料理念，并扩展到博物馆的所有公共空间。

Project name: Liangzhu Museum
Award date: 2008
Location: Liangzhu Cultural Village, China
Floor area: 9,500 m²
Architect: David Chipperfield Architects
Contact Architect: The Architectural Design and Research Institute at Zhejiang University of Technology
Client: Zhejiang Vanke Narada Real Estate Group Co., Ltd.
Photographer: Christian Richters, Chongfu Zhao
Completion date: 2007
Award name: McGraw-Hill Construction 2nd Bi-Annual "Good Design Is Good Business" China Awards 2008, Best Public Project

项目名称：良渚博物馆
获奖时间：2008
项目位置：中国 良渚文化村
建筑面积：9,500平方米
建筑设计：戴维·齐普菲尔德建筑师事务所
配合建筑：浙江工业大学建筑设计研究院
业主：浙江万科房地产集团有限公司
摄影师：Christian Richters 赵崇福
完成时间：2007
所获奖项：麦格劳-希尔公司《建筑实录》、《商业周刊》第二届"好设计创造好效益"中国奖项 2008最佳公共建筑

1

2

1. Entrance Courtyard
   入口前庭
2. Entrance Hall
   入口大厅
3. Exhibition
   展区
4. Central Courtyard
   中庭
5. Inner Courtyard
   内庭
6. Cafe
   休闲区
7. Shop
   商铺
8. Library / Office
   阅览室 / 办公室
9. Terrace
   平台

1. View from the South
   南侧视图

2. First Floor Plan
   一层平面图

3. View from the North
   北侧视图

4. Entrance Courtyard
   入口前庭

5. View from the Entrance Hall
   入口大厅视图

1. Inner Courtyard
   内庭空间

2. Public Area
   公共区

3. Central Courtyard
   中庭

3

1. Cafe Terrace
   休闲区平台
2. Conference Room
   会议室
3. Main Entrance
   主入口
4. View into the Inner Courtyard
   内庭视图
5. Main Entracce
   主入口

1

2

Dafen Art Museum

大芬美术馆

Dafen Gallery is located in Dafen Village of Shenzhen, which specializes in manufacturing and exporting handmade oil paintings. Their paintings are exported to Asia, Africa, Europe and America continents, creating several million *yuan* a year in sales even though they have long been regarded as a hybrid of vulgar art and business. The local government believes in the value of this creative industry, so Dafen Art Museum was decided to be built. While, this decision led to another question that whether this architecture could promote contemporary art and adjust the surrounding urban fabric as soon as it is completed. The design strategy is to design it into a comprehensive architecture that integrates the art gallery, commercial space, and rented studio into a whole. Several paths go through the whole building so as to enhance the communication of the people from different regions. Vertically, the gallery is located between the commercial space and public functions to allow the visual and spatial infiltration through varied functional areas. The exhibition, trade, painting and residence can co-exist within a building, thus weaving into a new form of the urban settlement. The gallery will inadvertently bury the value of the Dafen oil painting industry if it is too institutionalized, while to combine the gallery with the village is not only a form of reconciliation, but also a strategy to integrate the original ecology into the environment as well as the architecture.

位于深圳市龙岗区布吉镇的大芬村，是深圳著名的油画产业村，村中遍布油画复制品的创作坊。这里的油画出口到亚、非、欧、美各大洲，每年创造数亿元人民币的销售额。然而大芬村的油画长久以来是被视为一种低俗艺术，是庸俗品位与商业运作的奇妙混合体。但政府看到这种创意产业的价值，于是在一个似乎最不可能出现美术馆的地方，大芬美术馆出现了。这一决策引发了另一个问题，作为一项政府行为是否能在另一层面上促成当代艺术的介入，并且通过这一公众设施将周边的城市肌理进行调整，使日常生活、艺术活动与商业设施混合成新型的文化产业基地。设计策略是把美术馆、画廊、商业、可租用的工作室等等不同功能混合成一个整体，让几条步道穿越整座建筑物，使人们从周边的不同区域聚集于此，从而提供最大限度的交流机会。美术馆在垂直方向上被夹在商业和各种公共功能之间，并且允许在不同的使用功能之间有视觉和空间上的渗透。其结果是展览、交易、绘画和居住等多种活动可以同时在这座建筑的不同部位发生，各种不同的使用方式可以通过不断的渗透和交叠诱发出新的使用方式，并以此编织成崭新的城市聚落形式。以肯定大芬村油画产业为初衷的大芬美术馆的产生，如果机构化的气氛太浓，或许会不经意地埋葬大芬村油画产业。将美术馆与村落结合，既是一种形式上的调和，也是从本质上让自发的生活形态能在被设计的环境中得以延续和发展的策略。

Project name: Dafen Art Museum
Award date: 2008
Location: Shenzhen, China
Building area: 17,000 m²
Architect: Urbanus Architects & Design Inc.
Client: Longgang Municipal Government, Shenzhen
Completion date: 2007
Award name: McGraw-Hill Construction 2nd Bi-Annual "Good Design Is Good Business" China Awards 2008, Best Public Project

项目名称：大芬美术馆
获奖时间：2008
项目位置：中国 深圳
建筑面积：17,000平方米
建筑设计：都市实践设计事务所
业主：深圳龙岗区人民政府
完成时间：2007
所获奖项：麦格劳–希尔公司《建筑实录》、《商业周刊》第二届"好设计创造好效益"中国奖项　2008最佳公共建筑

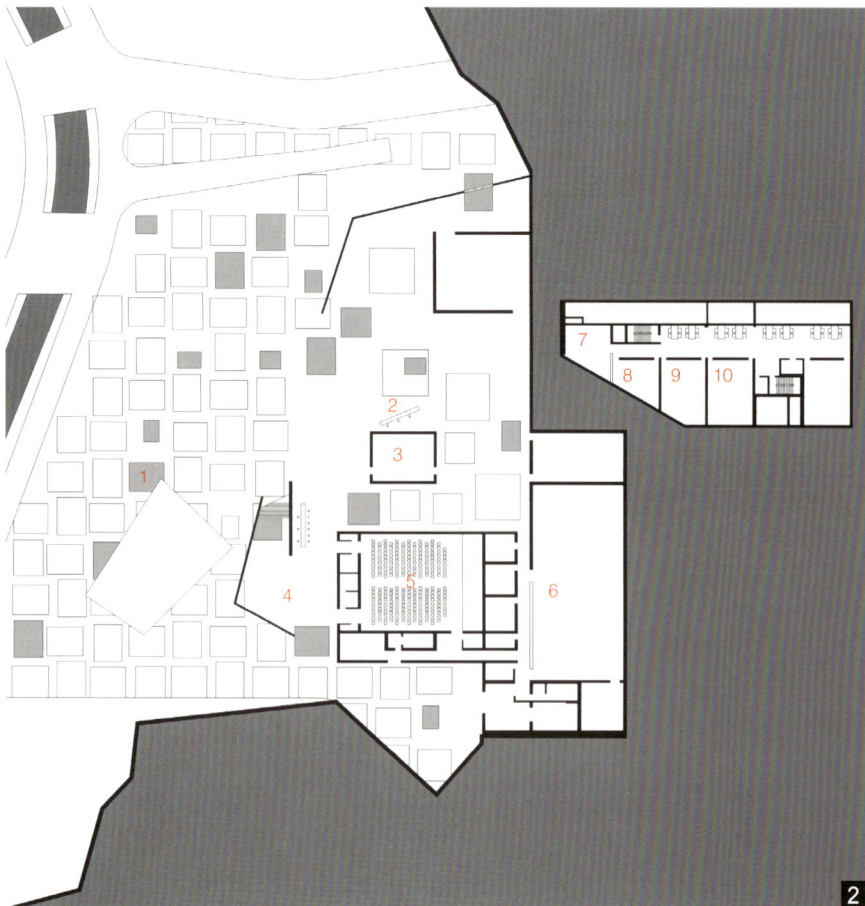

1. Outdoor Plaza
   室外广场
2. Reception Desk
   服务台
3. Guest Room
   洽谈室
4. Cafe
   咖啡休闲
5. Lecture Hall
   报告厅
6. Storage
   库房
7. Conference Room
   会议室
8. Office
   办公室
9. Office
   办公室
10. Manager
    馆长室

1. Facade
   正面图
2. Plan
   平面图
3. Bird View
   鸟瞰
4. Full View and Square
   全景及广场

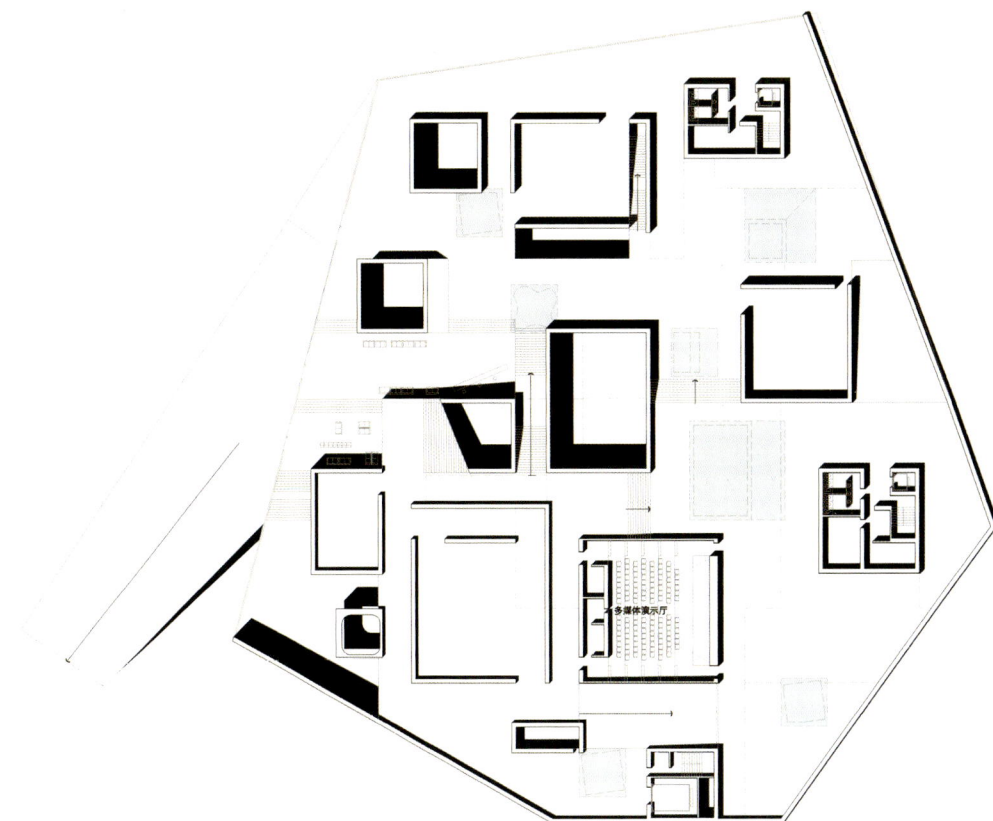

1. Interior
   室内

2. Display Area
   展区

3. Students in the Gallery
   画室里的学生们

4. Plan
   平面图

1. Rendering
   效果图
2. Rendering
   效果图
3. Exterior
   外墙

1

2

Suzhou Museum

苏州博物馆

The new Suzhou Museum is located in the northeast section of the historic quarter of Suzhou. It adjoins the landmarked Zhong Wang Fu, a complex of 19th-century historical structures, and the Garden of the Humble Administrator, a 16th-century garden listed as a UNESCO World Heritage site.

The design of the museum takes its cues from the rich vocabulary of Suzhou traditional architecture, with its whitewashed plaster walls, dark grey clay tile roofs and intricate garden architecture. However, these basic elements have been reinterpreted and synthesized into a new language and order, one that is contemporary and forward looking and hopefully one that is a possible direction for the future of Chinese modern architecture.

As with traditional Suzhou architecture, the design of the Art Museum is organized around a series of gardens and courts that mediates between the building and its surrounding environment. The main Museum Garden is a contemporary extension and commentary of the Garden of the Humble Administrator to the north. As visual connections between the two properties are not possible due to the high garden walls, water is used physically and metaphorically as a bridge between the two properties.

The new Museum Garden and its smaller Gallery and Administrative Gardens are not landscaped based on traditional and conventional approaches. Rather, new design directions and themes have been sought for each of them, where the essence of traditional landscape design can be distilled and reformulated into potentially new directions for Chinese garden architecture.

苏州博物馆新馆位于苏州古城历史保护区东北部，毗邻建于清代的全国重点文物保护忠王府，以及被联合国教科文组织列为世界文化遗产的明代名园拙政园。

博物馆的建筑设计借鉴了苏州传统建筑艺术的表现方式，承袭了粉墙黛瓦的建筑风格和精致的园林布局艺术，并赋予新的含义，以求"中而新、苏而新"。通过传统与创新的结合，试图寻求中国现代建筑艺术发展的方向。

如同苏州传统园林建筑，苏州博物馆的建筑与庭园融为一体，且相辅相成。博物馆主庭园由水池、假山与造型别致的茶亭组成。主庭园与拙政园一墙相隔，其简洁的园林创作风格可谓传统园林的现代版诠释，"以壁为纸、以石为绘"的片石假山以宋代米芾山水画为蓝本，俨然是一幅立体的山水画。

Project name: Suzhou Museum
Award date: 2008
Location: Suzhou, China
Building area: 17,000 m²
Architect: I. M. Pei Architect with Pei Partnership Architects and Suzhou Institute of Architectural Design
Client: Suzhou Municipal Administration of Culture, Radio& Television
Photographer: Kerun Ip
Completion date: 2006
Award name: McGraw-Hill Construction 2nd Bi-Annual "Good Design Is Good Business" China Awards 2008, Best Public Project

项目名称：苏州博物馆
获奖时间：2008
项目位置：中国 苏州
建筑面积：17,000平方米
建筑设计：贝聿铭建筑师与贝氏建筑事务所   苏州市建筑设计研究院
业主：苏州市文化广播电视管理局
摄影师：Kerun Ip
完成时间：2006
所获奖项：麦格劳-希尔公司《建筑实录》、《商业周刊》第二届"好设计创造好效益"中国奖项   2008最佳公共建筑

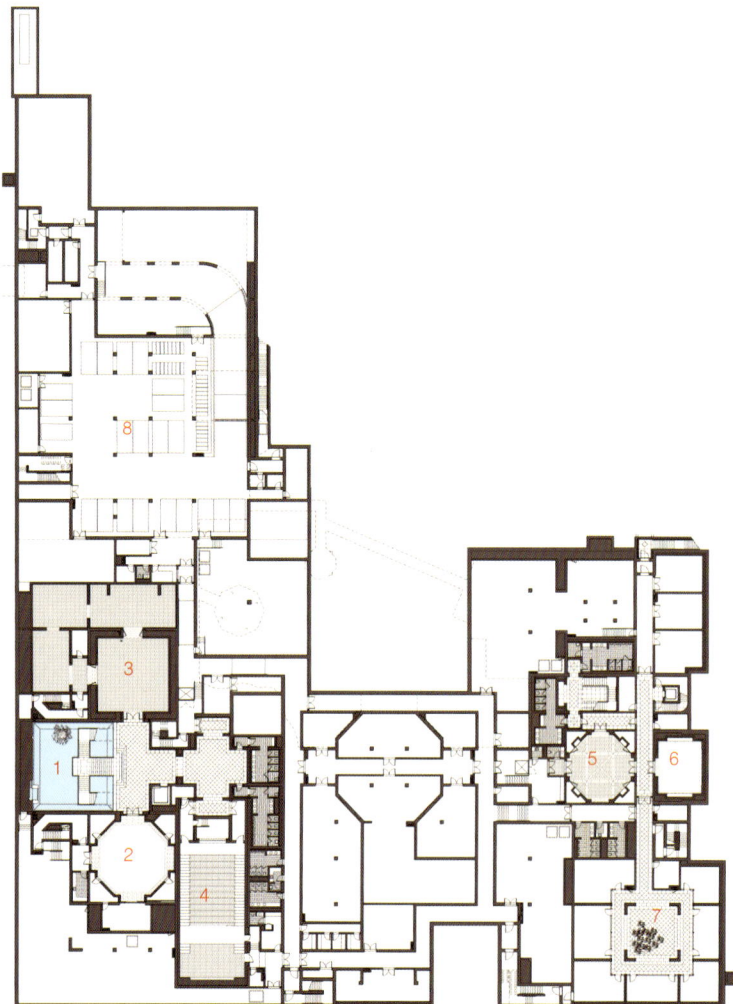

1. Lotus Pool
   莲花池
2. Orientation Hall
   多功能厅
3. Special Exhibitions
   特展厅
4. Auditorium
   学术报告厅
5. VIP Viewing Room
   贵宾观摩厅
6. Conference Room
   会议室
7. Offices
   办公区
8. Carpark
   车库

1. Garden Bridge
   庭院小桥
2. Plan
   平面图
3. Entrance Court
   入口前庭
4. Main Entrance
   正门

1

2

1. Rock Landscape
   石刻景观

2. Courtyard
   庭院

3. Contemporary Art Gallery
   现代艺术展厅

4. Porcelain Gallery
   瓷器展厅

1.Great Hall
正厅
2.Painting Gallery
绘画展厅

2

Shanghai South Station

上海铁路南客站

An international competition for the Shanghai South Station took place from May to September 2001, over 3 different phases. AREP and ECADI won this competition jointly. With its design, the Shanghai South Station goes beyond the traditional function of the railway station as a mere exchange hub, to become an active participant in the role of Shanghai as an economic, financial, cultural and creative world-class city. A destination in itself, the station provides travellers and visitors alike many services as well as prestigious retail spaces. Its circular form allows for a simple and straightforward operational mode: it provides good circulation conditions to vehicles, and it offers travelers the most direct and logic way to waiting rooms and platforms.

The most striking architectural element is the 255 m diameter roof, which, despite an area of 60,000 m², still remains a light, slender and elegant element of the project. It is composed of three different layers: the external sun breakers, the transparent polycarbonate surface and the internal perforated metal skin. The combination of those three layers filters and diffuses natural light throughout the space.

The lighting principles of the station constitute an essential design element. Light poles are distributed evenly over the station according to a rigorous and precise grid, in order to obtain a homogeneous light. The use of concealed lighting, lamps trained directly on the under-face of slabs and roofs, guarantees that the station remains a bright beacon at night. Using materials such as glass (transparent or pattern printed), transparent polycarbonate elements, perforated steel or brushed aluminium to achieve a play of transparency, gives the station both lightness and a modern expression. Urban, regional, as well as long distance means of transportation all converge towards Shanghai South Station, which in turn becomes more than a simple hub, and fully assumes its role as the gate to the city.

上海铁路南客站的国际竞标从2001年5月持续到9月，经历了三个阶段，最终由AREP和ECADI共同获胜中标。这个设计，使上海南客站远远超越了一个传统火车站单纯的交换场所的角色，而是成为了一个可以以经济、金融、文化和创造性角色融入上海的积极参与者。就本身而言，车站为旅客和访客提供了包括零售名店的许多服务。它的圆环形设计考虑到了一种既简单又直接的操作模式：既为机动车提供良好的交通，又为旅客提供方便的候车和登车路径。

设计中最震撼的元素就是250米直径的屋顶，尽管占地60000平方米，也依然使整个项目光感通透，尽显细致和典雅的姿态。它由三个不同的层次构成：最外面的防光层、中间透明的聚碳酸酯纤维材料和内部打孔的金属层。这三层结构的组合可以过滤和散射自然光。

车站的照明系统是整个设计的关键性元素。为了得到均匀和谐的光感，根据严格精确的测量，灯柱被均匀的分布在车站周围。项目使用了隐形的照明设备，即灯光被安置在棚顶和周围，保证即使在夜间，整个车站也灯火通明。建筑所使用的材料，比如玻璃（透明的或有花纹的）、透明聚碳酸酯纤维材料、穿孔的钢或者铝，都为上海南客站赋予了一种现代气息。城市气息、地方气息和各种长途交通工具都在上海南客站汇集，使这个项目超越单纯的交通场所，成为了承担更多角色的上海门户。

Project name: Shanghai South Station
Award date: 2008
Location: Shanghai, China
Site area: 47,000 m²
Architect: AREP VILLE, East China Architectural Design&Research Institute Co., Ltd. / Shanghai Xian Dai Architectural Design Group
Client: the Shanghai Railway Station Bureau
Photographer: D. Boy De La Tour, T. Chapuis
Completion date: 2006
Award name: McGraw-Hill Construction 2nd Bi-Annual "Good Design Is Good Business" China Awards 2008, Best Public Project

项目名称：上海铁路南客站
获奖时间：2008
项目位置：中国上海
占地面积：47,000平方米
建筑设计：法国AREP 现代设计集团华东建筑设计研究院
业主：上海铁路管理局
摄影师：D. Boy De La Tour T. Chapuis
完成时间：2006
所获奖项：麦格劳-希尔公司《建筑实录》、《商业周刊》第二届"好设计创造好效益"中国奖项 2008最佳公共建筑

1. Night View
   夜景

2. Exterior
   外观图

3. Retails
   商铺

1

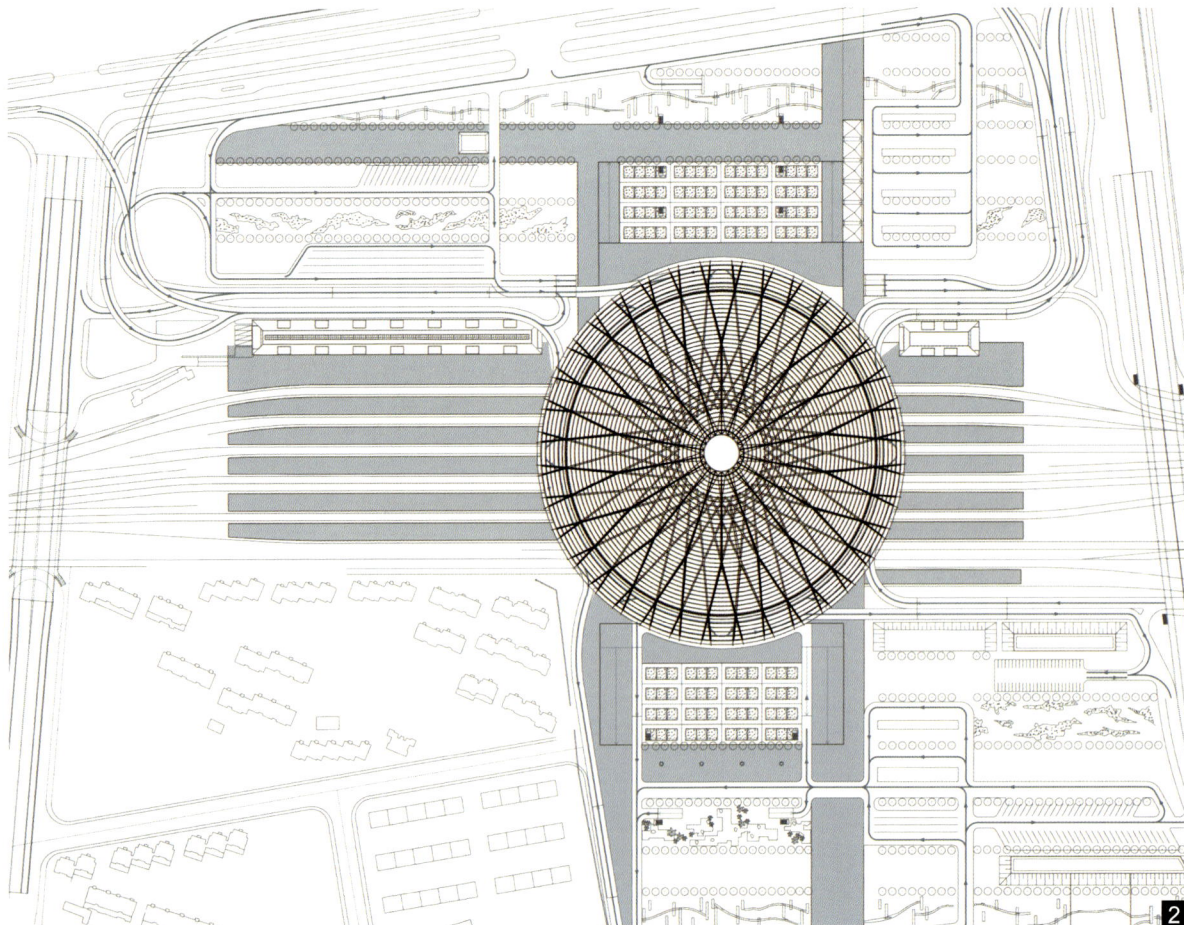

2

1. Waiting Area
   候车区

2. Plan
   平面图

3. Escalator
   电梯

4. Entrance
   入口

1. Entrance from Northeast
   东北入口

2. Exterior
   外观

3. Landscape
   外部景观

2

Sino-French Center, Tongji University

同济大学中法中心

Sino-French Center of Tongji University is located at the south-east corner of the campus, with 12.9 Building, the oldest existing building of the campus, and 12.9 Memorial Park on its west side, tracking field on its south side, and Siping Road on its east side. XuRi Building, which should be preserved, is located at the north-west corner of the site.

The goal of this project is to create a form system to integrate its program, its site context and its culture context. The designer achieved this by using a geometric diagram to control the materialization of its program and circulation, to conform to the site restriction, and also to indicate its symbolic meaning, the culture exchange between two countries. This diagram of "Hand in Hand" is introduced to organize the whole building with its inherent structure of dualistic juxtaposition.

The program is composed of three parts, college, office and public gathering space. Two similar but different zigzag volumes, occupied by college and office sector respectively, overlap and interlace each other, and then they are linked together by the volume of public gathering space on underground and up level. The function of the college and offices is well kept in mind by using regular shapes for almost each unit. Yet applying zigzag corridor to connect these units creates abundant interests throughout inside and outside space. In the meanwhile, existing trees are incorporated into the design to add more charms to this complex.

Different materials and tectonics are applied to the different components of the complex. The college sector is wrapped by COR-TEN Steel sheet panels. The unique texture and color of the panels and the smoothness of the glass create a delicate variation. Precoated cement panel is introduced into the office sector. Regular and irregular window bands provide sunlight to the office units and corridors. Public gathering space is created by the combination of both COR-TEN steel panel and precoated cement panel. The vivid color and texture of COR-TEN steel panel is contracted with plain grey cement panel. This treatment indicates the symbolic meaning of this project, the juxtaposition of two different cultures.

同济大学中法中心位于校园东南角，西临校园内现存最老的建筑物一二、九大楼和一二、九纪念园，南侧为运动场，东侧紧靠四平路。

把这一建筑看作建筑形式系统对内部使用功能和外部环境条件及更广阔的文化语境的创造性整合。从项目本身所具有的多层面的"交流性"入手，提出了一个"双手相握"的图解，利用这一图解的潜在"二元并置"结构来组织整个建筑的相关系统，以达成一个"和而不同"的整体。这个图解既是对建筑内部功能和流线系统的抽象，又源于场地条件对建筑体量的挤压和拉伸，同时也是对中法两国文化的差异存的关照。

整个建筑分为既分又合的三个部分，分别用于教学、办公和公共交流。南北两条进深相同、由曲折连廊串联大小使用空间的教学、办公单元互相穿插后分别从空中和地下结合到最北端的公共交流单元。教学、办公单元的共用门厅位于它们上下穿插的虚空部分，通透高耸，强化了两者的穿插关系。公共交流单元另设一个独用门厅，并将地下的展厅、南侧的屋顶水池、下沉庭院和二层的报告厅联系起来。

三个不同单元采用不同的材质组合、色彩和构造做法来建构。教学单元用自然氧化的耐候钢板包裹网格状立面，均质的网格中开孔和玻璃微妙地变化；办公单元用无机预涂装水泥纤维板覆盖立面，规则条窗和不规则条窗分别为办公室和走廊提供光线；公共交流单元是无机预涂装水泥纤维板和耐候钢板的混合立面，外表皮为无机预涂装水泥纤维板，大尺度开口部位为耐候钢板。这样的两种色彩和材质暗示了中法不同的文化传承的视觉表征。

Project name: Sino-French Center, Tongji University
Award date: 2008
Location: Shanghai, China
Gross Floor Area: 13,575 m²
Architect: Atelier Z+
Client: Tongji University
Photographer: Zhang Siye
Completion date: 2006
Award name: McGraw-Hill Construction 2nd Bi-Annual "Good Design Is Good Business" China Awards 2008, Best Public Project

项目名称：同济大学中法中心
获奖时间：2008
项目位置：中国 上海
建筑面积：13,575平方米
建筑设计：致正建筑工作室
业主：同济大学
摄影师：张嗣烨
完成时间：2006
所获奖项：麦格劳–希尔公司《建筑实录》、《商业周刊》第二届"好设计创造好效益"中国奖项　2008最佳公共建筑

1

2

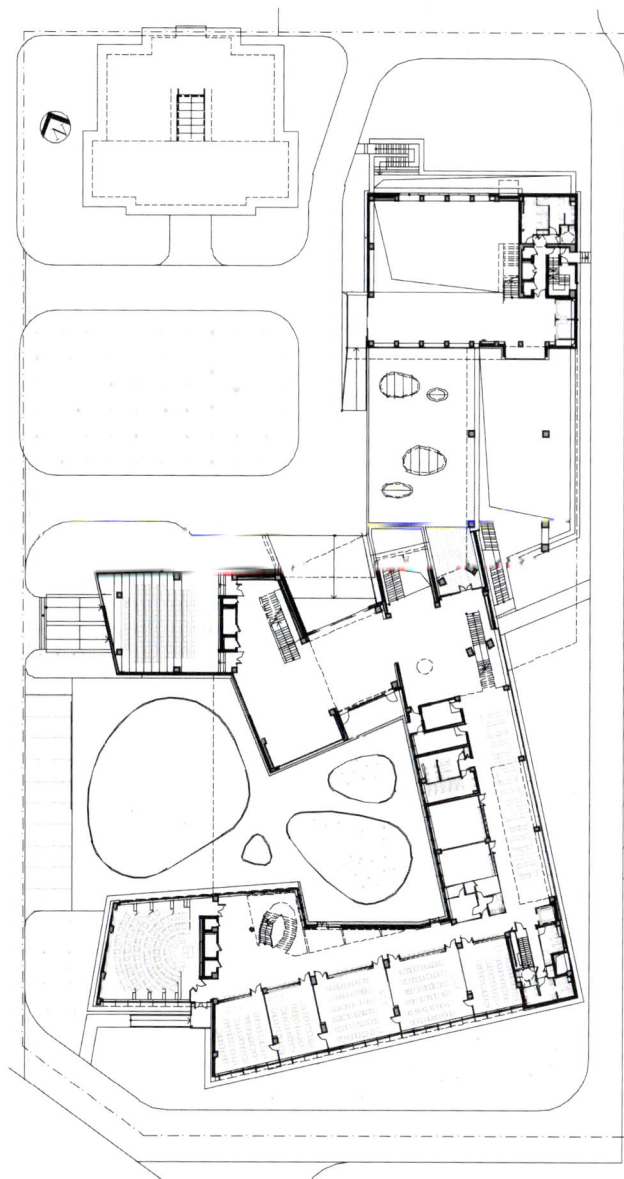

3

4

1. Stair of College Sector
   教学单元旋转楼梯

2. Lecture Hall
   报告厅

3. Plan
   平面图

4. Interior of College Sector
   教学单元教室局部

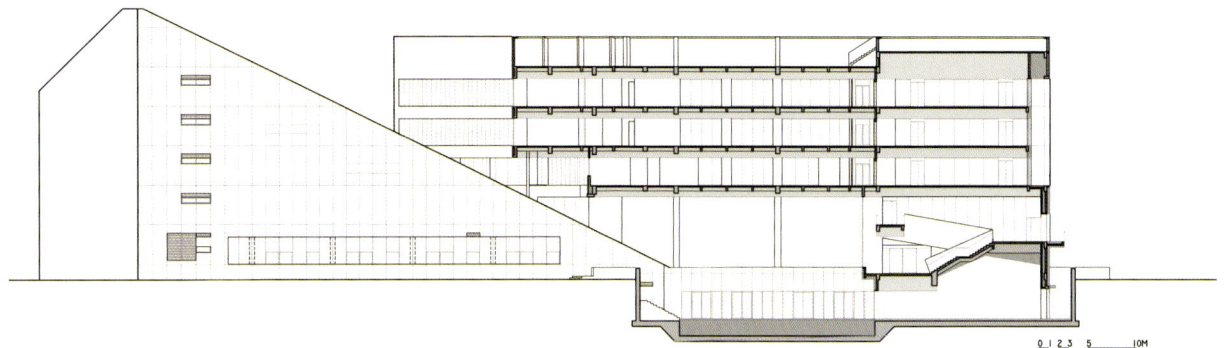

1. View from West
   西侧局部

2. View from North
   北侧局部

3. Section
   剖面图

1. View from Southwest
西南側局部

2. Sunk Garden
下沉庭院

2

Luyeyuan Stone Sculpture Art Museum

鹿野苑石刻艺术博物馆

The site of this project is a plain field between the riverbed and the woods, composed of mainly four areas. On the largest one is located the major part of the museum, while the other three are arranged respectively for parking, open exhibition and subordinate building. Bamboos become the natural divisions of each part. Routes link different areas, gradually float upon the ground, through the bamboos and finally lead to the entrance above the lotus pond. Museum display is designed around an atrium, and the light, exhibits and the landscape are organized by carefully dealing with gaps between building blocks.

The museum is a collection of stone sculptures, thus the architecture wants to tell people a story of "Artificial Stone". As local construction technique is low which makes flexible modification afterwards possible, a combined technique of frame structure with fair-faced concrete and shale bricks is created. Bricks on the inner side of the combined wall are used as a template to make sure that the concrete is poured vertically, as well as to serve the later modifications as a soft liner. Indention templates with stripes are used to pour the fair-faced concrete walls of the main building in order that a clear pattern is formed, a strong impression of walls is made and the defect caused by the lack of experience on the pouring technique is hidden behind the dense and wild lattice. All these above are adopted to satisfy aesthetic and spiritual pursuits of the architect, at the same time, to solve various problems China today is being faced.

In China, due to the inconsiderate aforehand plan, architectural projects are more likely to change optionally afterwards. The interior of the mixed wall adopts bricklaying and plastering technique to cope with different kinds of changes afterwards.

Because of the rude construction technology, it's very hard to assure the perpendicularity of the walls during the pouring course. If laying is done first, the interior walls can be the formwork when pouring the outer walls, and thus the perpendicularity is ensured.

The designer hopes to find out a way, an approach to contemporary architectural aesthetic ideals while feasible and proper in local condition.

博物馆用地是河滩与竹林相间的平地。主体位于其中最大的一块林间空地上。其余三块空地分别为前区停车场、露天展区，后勤附属用房。竹林成为其间的自然分隔。路径串连起各区域，沿途逐渐架起，临空穿越竹林并引向莲池上的入口。博物馆采用展厅环绕中庭的布局，利用建筑体块之间的间隙组织光线、展品和风景。

博物馆藏品以石刻为主题，建筑则希望表现一部"人造石"的故事。针对当地低下的施工技术以及事后改动随意性极大的情况，采用 "框架结构、清水混凝土与页岩砖组合墙"这一特殊的混成工艺，利用组合墙内层的砖作为内模以保证混凝土浇筑的垂直度，同时成为"软衬"以应付事后的开槽改动等。整个主体部分清水混凝土外壁采用凸凹窄条模板，一是为了形成明确的肌理，增加外墙的质感和可读性，同时，粗犷而较细小的分格可以掩饰由于浇筑工艺生疏而可能带来的瑕疵。希望既满足建筑追求又解决中国的问题。

中国的建筑项目通常事前策划不严密，事后改动随意性极大。组合墙的内壁采用砖砌抹灰，可以应付开槽、埋线、装配挂钩支架等事后改动。

由于施工技术原始，难以在浇筑过程中保证墙体的垂直度，利用先砌内壁后浇外层的作法，可以将内壁墙体作为模板，易于保证垂直度。

设计师希望寻找到一种方法，它既在当地是现实可行，自然恰当的，又能够真实地接近当代的建筑美学理想。

Project name: Luyeyuan Stone Sculpture Art Museum
Award date: 2006
Location: Sichuan, China
Floor Area: principle part 990 m²
Architect: Jiakun Architect & Associates
Client: Labor Organization of Xiangcai Securities
Photographer: Bi Kejian
Completion date: 2002
Award name: McGraw-Hill Construction 1st Bi-Annual "Good Design Is Good Business" China Awards 2006, Best Public Project

项目名称：鹿野苑石刻艺术博物馆
获奖时间：2006
项目位置：中国 四川
建筑面积：主体990平方米
建筑设计：家琨建筑设计事务所
业主：湘财证券工会
摄影师：毕克俭
完成时间：2002
所获奖项：麦格劳–希尔公司《建筑实录》、《商业周刊》第一届"好设计创造好效益"中国奖项 2006最佳公共建筑

1

2

1. Main Entrance
   主入口

2. Main Entrance
   主入口

3. Courtyard
   庭院

4. Plan
   平面图

3

1. Lotus Pool
　莲花池
2. Exhibition Room
　展厅
3. Atrium
　中庭
4. Pool
　池塘
5. Multi-function Hall
　多功能厅
6. Office
　办公室

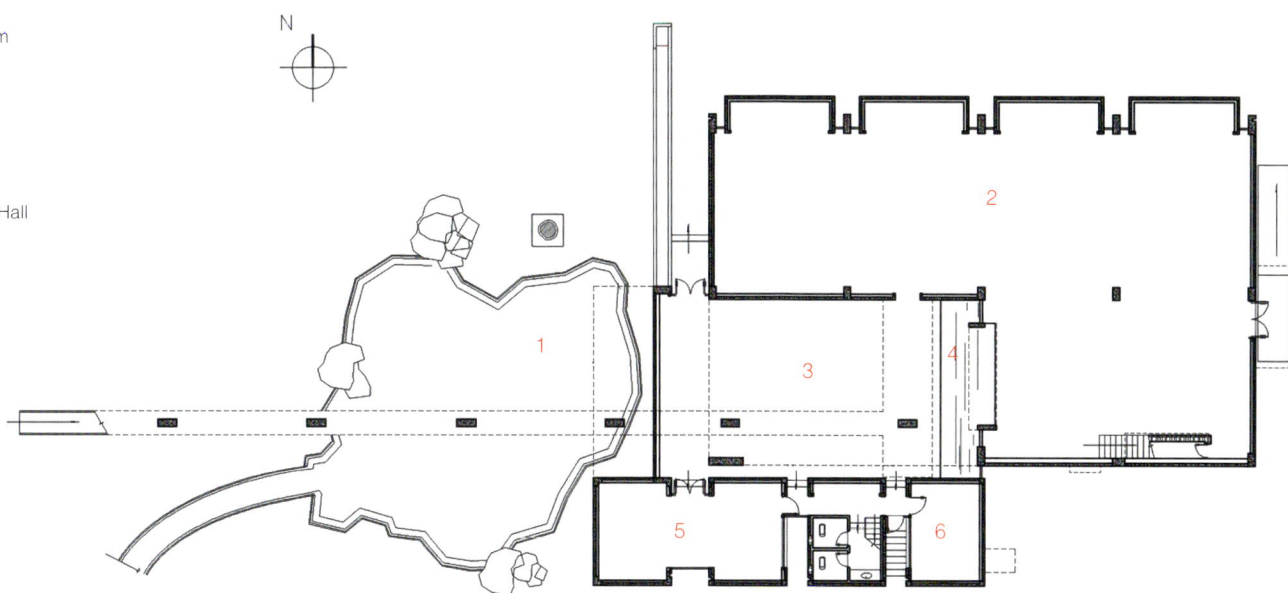

4

1. Roof
   屋顶

2. Courtyard
   庭院

3. Interior
   室内

4. Interior
   室内

Shenzhen Planning Building

深圳市国土局办公楼

The flexible office of the Shenzhen Planning and Land Resource Bureau has declared its vector architecture – Shenzhen Planning Building should have a non-bureaucratized image. Therefore, the interior and the exterior of the building are in equal height, without a transitional high step; its internal space is connected with the city, rather than hidden in a mysterious hole. The most destructive factor in the modern urban development is the institutionalization that the individual or small group of institutional building plays the leading role in the main public buildings in the city but disliked by the public. Their model has essentially destructed the overall city space. The Folunluosa Cathedral that shelters all of the Tuscan people is always loved by the public, while, people will no longer to see such buildings now because the cities were mutilated by the institutionalized building. Thus, in the Planning Building hall, a peaceful, natural, open interior space is expected, and even the people can freely have a break here (a passive ventilation design makes it possible). The office design is based on a sustainable development model. The office unit constitutes an office system. Each unit is a double room, dependent on a larger office space. The structure of light steel beam floor can keep up with the space needs.

弹性办公——深圳规划与国土资源局的窗口式办公方法，已经宣告了它的建筑载体——深圳规划与国土资源局办公楼——所应有的非衙门化形象。因此，这个建筑的地平与室外是相平的，而不是通过高台阶的转折；其内部与城市是相通的，而不是隐蔽在神秘窗洞里中。近代城市发展中的最具有破坏性的因素是机构化，即属于个体或小团体的机构性建筑成为城市公共建筑的主体，却不能被整个市民所享受。因此，它们的造型从本质上是对城市整体空间的挑战与破坏，人们再也看不到像佛伦罗萨大教堂那样拥有市民共同价值的建筑，那种能够把所有塔斯干人都庇护住的建筑，因而，城市被机构化的建筑肢解了。而在国土局的大厅里，我们希望看到的是一种和平、自然、开放的室内空间，甚至有一天市民们都能自由来到这里乘凉（一种被动式通风的设计使之成为可能）。办公模式建立在一种可持续发展的模式基础之上，以办公单元构成一个办公系统。每一个单元是一个复式空间，它依存于一次性构筑的大办公室空间。其结构为轻钢梁柱楼板体系，可以随空间需要不断增长。

Project name: Shenzhen Planning Building
Award date: 2006
Location: Shenzhen, China
Total area: 33,400 m²
Architect: Urbanus
Client: Shenzhen Urban Planning Bureau
Completion date: 2005
Award name: McGraw-Hill Construction 1st Bi-Annual "Good Design Is Good Business" China Awards 2006, Best Public Project

项目名称：深圳市国土局办公楼
获奖时间：2006
项目位置：中国深圳
总面积：33,400平方米
建筑设计：都市实践建筑设计事务所
业主：深圳市国土局
完成时间：2005
所获奖项：麦格劳-希尔公司《建筑实录》、《商业周刊》第一届"好设计创造好效益"中国奖项　2006最佳公共建筑

1. Great Hall
   正厅

2. Interior Pathway
   室内通道

3. Part of Exterior
   外墙一角

3

1. Outside Scenery through
Windows
窗户透视户外
2. Plans
平面图
3. Interior
室内
4. Light and Shadow
光影效果
5. Stairs
楼梯

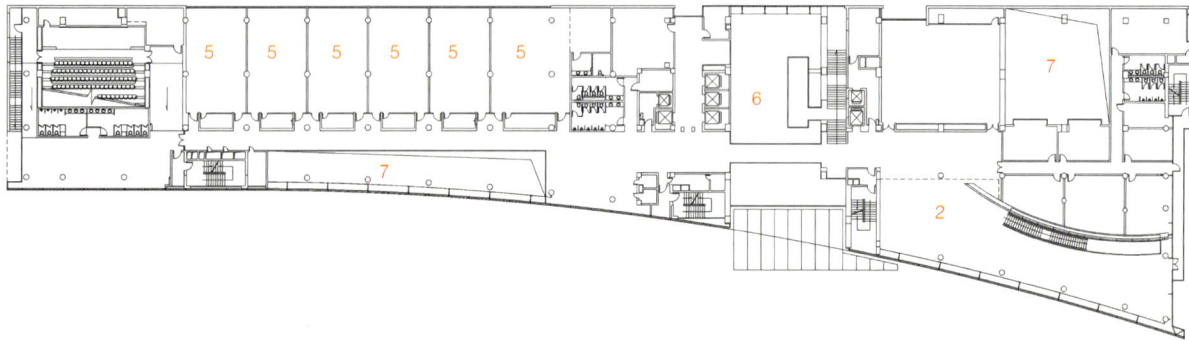

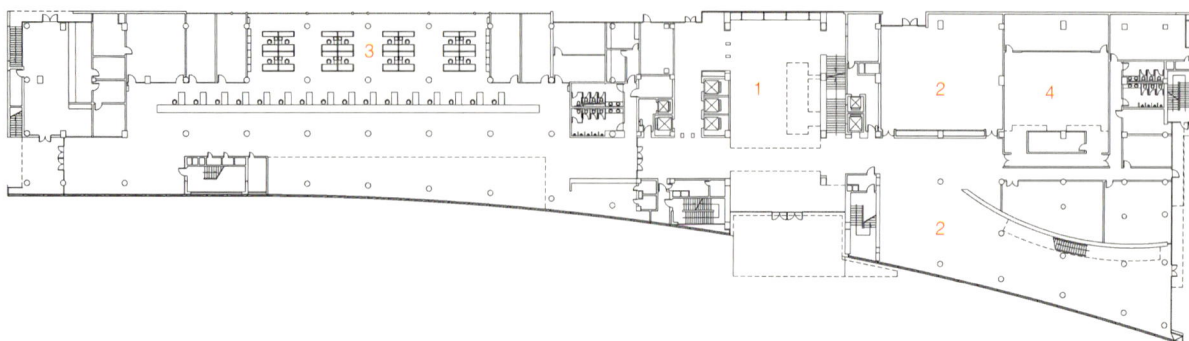

1. Entrance Hall
入口大厅
2. Exhibition
展区
3. Window Office
办公窗口
4. Animation Hall
生态大厅
5. Conference Room
会议室
6. Info-box
信息盒子
7. Atrium
中庭

Yuhu Elementary School and Community Center in Lijiang

玉湖小学

Yuhu is a Naxi minority village in the UNESCO World Heritage site of Lijiang, ten hours' bus ride North-west of Kunming. On the foothills of the Jade Dragon Snow Mountain, it is 2,760 meters above sea level and so enjoys a cool-dry climate with cool summer and mild winters. Behind it rise the white peaks of the mountains, dramatically defining the horizon.

With the Naxi tenet of the mountains as the backbone and water as the soul of their culture, the design incorporated local stones and water. The most plentiful material in the area is white-colored sedimentary limestone as well as cobblestones. Considering of material sustainability, the architects decided to use them as much as possible, cobblestones for water features and limestone for the rest. Big sliding and casement fenestrations open in clement weather allow as much light to penetrate as possible. All traditional aesthetic treatments or ornamentations were reduced to their basics: curved roof ridges were straightened and gable end ornamentations simplified into a timber lattice frame inspired by traditional grains racks.

Traditional construction materials and techniques have been carefully married to modern ones. Timber was brought, cut and planed on site to make the structural frame and fenestrations. A structure design challenge was how to allow the buildings to withstand lateral loading in this earthquake zone.

Unlike traditional Naxi houses that sit on stone pad foundations, reinforced concrete pad foundation with ground beams was used. Stonewalls were also reinforced with vertical rebars and horizontal wire mesh at regular intervals to resist lateral forces during an earthquake. While the concrete may have been relatively new, compared to traditional mud/lime mortar, the stones and timber were prepared in traditional fashion. New ways of composing and putting together traditional materials create interesting juxtaposition of new and old, reinventing the traditional Naxi house for modern usage.

玉湖小学的设计目的是结合教学、科研、设计与建造实践, 从文化、经济、自然资源及环境角度出发, 为世界文化遗产所在地提供一个可持续发展的范例。

李晓东教授的设计理念的产生建立在他对当地传统、建造技术、建筑材料以及资源的研究基础之上。所有的传统审美处理手法还有装饰都被简约化: 弯曲的屋顶曲线被拉直, 山墙的装饰也被简化成木制格栅, 其灵感来源于传统的农家晾谷架。乡土建筑的精华被提炼, 并以形式和空间的方式表达出来。这里, 建筑师保留了坡屋顶的基本元素, 例如灰色瓦砖的运用, 以及将内部空间以传统的正开间划分等等。

基于以山为骨、以水为魄的纳西文化, 设计者有意识地将当地材料和元素最大程度地运用到了设计中。并且对于材料的可持续发展的考虑, 在设计中大量采用了当地资源丰富的白色石灰沉积岩和卵石, 主要使用在石墙和铺地上。由于周围的建筑大多是由粘土砖建成, 白色石灰石的运用使得该建筑即卓尔不群, 又显得很和谐。

Project name: Yuhu Elementary School and Community Center in Lijiang
Award date: 2006
Location: Lijiang, China
Floor area: 830 m²
Architect: Li Xiaodong Atelier
Client: Yuhu Village
Photographer: Melvin H J Tan
Completion date: 2004
Award name: McGraw-Hill Construction 1st Bi-Annual "Good Design Is Good Business" China Awards 2006, Best Public Project

项目名称: 玉湖小学
获奖时间: 2006
项目位置: 中国 丽江
建筑面积: 830平方米
建筑设计: 李晓东工作室
业主: 玉湖村
摄影师: Melvin H J Tan
完成时间: 2004
所获奖项: 麦格劳-希尔公司《建筑实录》、《商业周刊》第一届"好设计创造好效益"中国奖项 2006最佳公共建筑

1. Courtyard
   院落

2. Stairs
   楼梯

3. Classrooms
   教室

4. Plans
   平面图

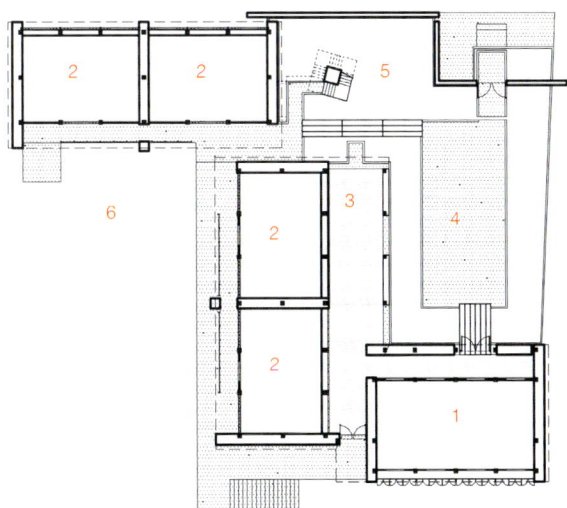

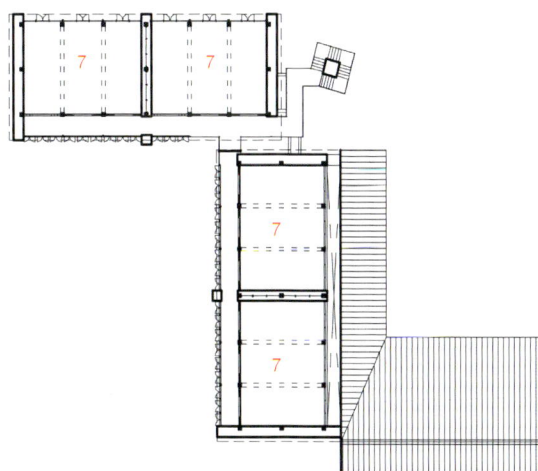

3

4

1. Museum
博物馆

2. Classroom
教室

3. Exhibition Area
展区

4. Community Courtyard
社区庭院

5. Reflecting Pool
倒映池塘

6. School Courtyard
学校庭院

7. Staff Room
员工休息室

1. Courtyard
院落

2. Water
水景

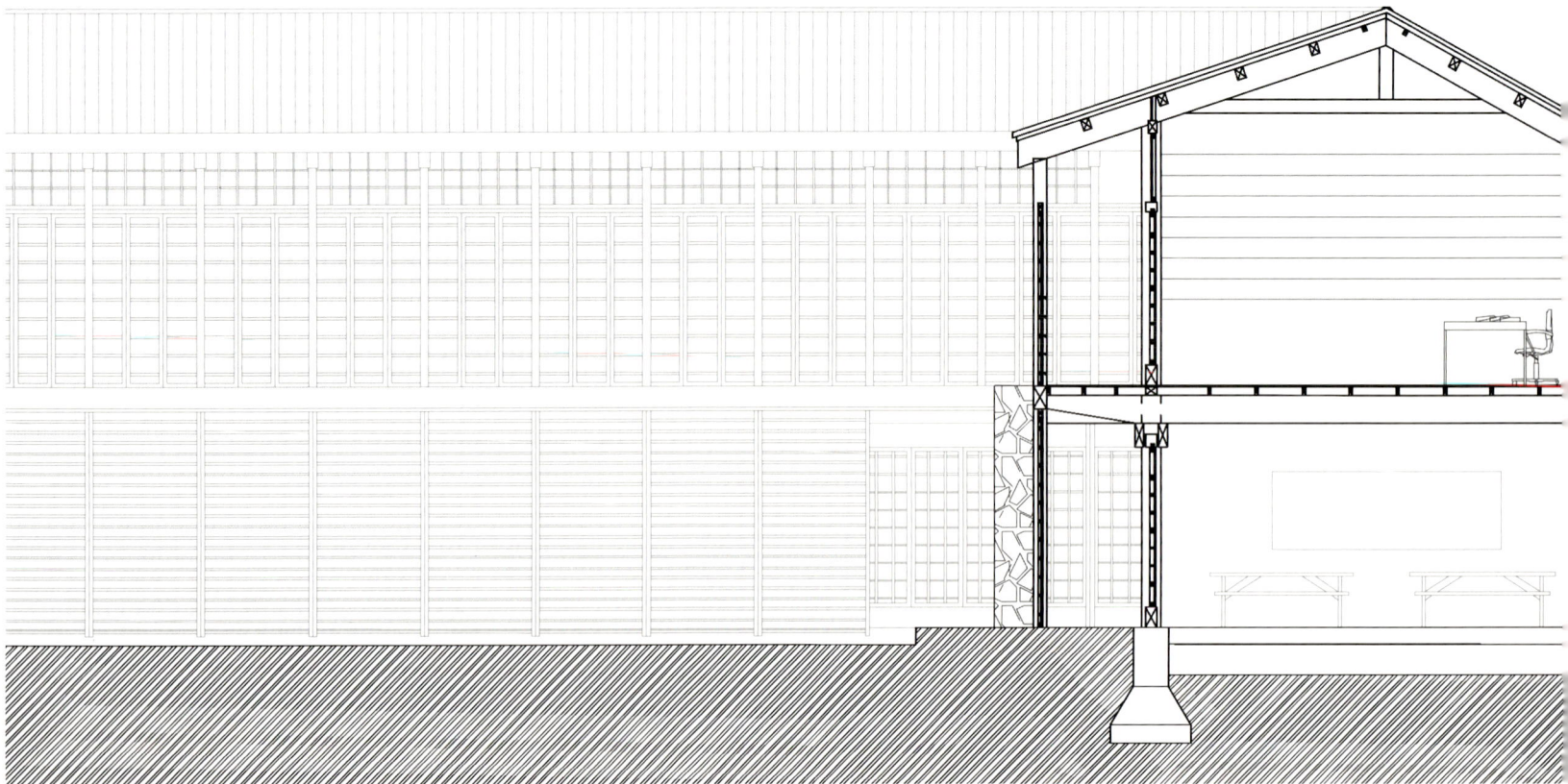

1. Resting Room for Teachers
   教师休息室
2. Section
   剖面图

Evian Town

依云水岸

Evian Town is a low-rise, high-density villa community. The scheme combines the traditional essence of Suzhou with modern architecture, with the overall building and landscape designed to respect a human scale. It comprises intimate residential clusters, meandering pedestrian spaces, subtle materials and a unique regionally inspired color palette.

The scheme's layout follows a randomly staggered order with interconnected living spaces that contribute to creating a living environment with a community feel. In keeping with the traditional Suzhou residential style, the villas have courtyards that are used as a "green threshold" between the living and dining areas, creating an energetic flow and allowing ventilation into all spaces.

The fundamental order within the masterplan is derived from creating intimately scaled residential clusters woven together with a series of interconnected pedestrian routes and shared public spaces. The intention is to create a sense of neighborhood and "ownership" within the greater residential fabric.

Front and rear gardens enclose each villa providing some privacy. Full height glass windows connect the green spaces to the interior living quarters while giving them more natural light as well as an impression of enhanced spaces.

依云水岸是一个融合了苏州传统建筑精髓的大型现代低密度别墅区。RMJM对该项目的设计理念是运用人性化尺度的公共空间，小巧的建筑体量，细腻的材料选择及色彩搭配。营造出具有中国江南建筑内蕴的居住环境，打破一般现代建筑物冰冷巨大的感觉。并配合带有苏州传统园林独有布局来营造不一样住宅精品的感觉，唤醒人们心中的文化应同感。

在住宅区设计中，别墅的布局自由错落分布，高低有致，形成有机的社区居住环境。为了保留传统的苏州居住特色，每幢别墅均内设中心庭院，成为客厅与饭厅之间的绿色借景，创造出流动的空间感，并为所有内部空间带来了自然光线和通风。

总体规划创造柳暗花明的行人小径和公共空间，与私密住宅小区紧密交织，在大型住宅社区中保持邻里空间的归属感。

别墅前后也设置了庭院，一方面提供了户外过渡空间以确保用户的私密性；而另一方面利用大面积的落地玻璃把室外庭院的景观引入室内，打造出独特和谐的居住环境。

Project name: Evian Town
Award date: 2006 - 2010
Location: Suzhou, China
Building area: GFA approximately 271,000 m²
Designer: RMJM Hong Kong Limited
Local Design Institute: Suzhou Institute of Architectural Design Co. Limited (SIAD)
Huasen Institute of Architectural Design & Research Co., Ltd
Client: China Merchants Real Estate (Suzhou) Ltd.
Key Designers: Scott Findley, Dimi Lee, Rita Pang
Photographer: Jason Findley, Hans-Georg Esch, But-Sou Lai
Completion date: 2009
Award name: McGraw-Hill Construction 3rd Bi-Annual "Good Design Is Good Business" China Awards 2010, Best Residential Project

项目名称：依云水岸
获奖时间：2006 – 2010
项目位置：中国 苏州
建筑面积：约 271,000 平方米
主设计单位：罗麦庄马香港有限公司
本地设计院：苏州市建筑设计研究院有限责任公司
华森建筑与工程设计顾问有限公司
业主：招商局地产（苏州）有限公司
主设计师：Scott Findley, 李志浩, 彭凯欣
摄影师：Jason Findley,Hans-Georg Esch, 黎不修
完成时间：2009
所获奖项：麦格劳–希尔公司《建筑实录》、《商业周刊》第三届"好设计创造好效益"中国奖项　2010最佳住宅建筑

Evian Town Phase III
依云水岸三期

Photography: Jason Findley

1. Clubhouse
会所

2. Courtyard
内园

3. Courtyard
内园

1

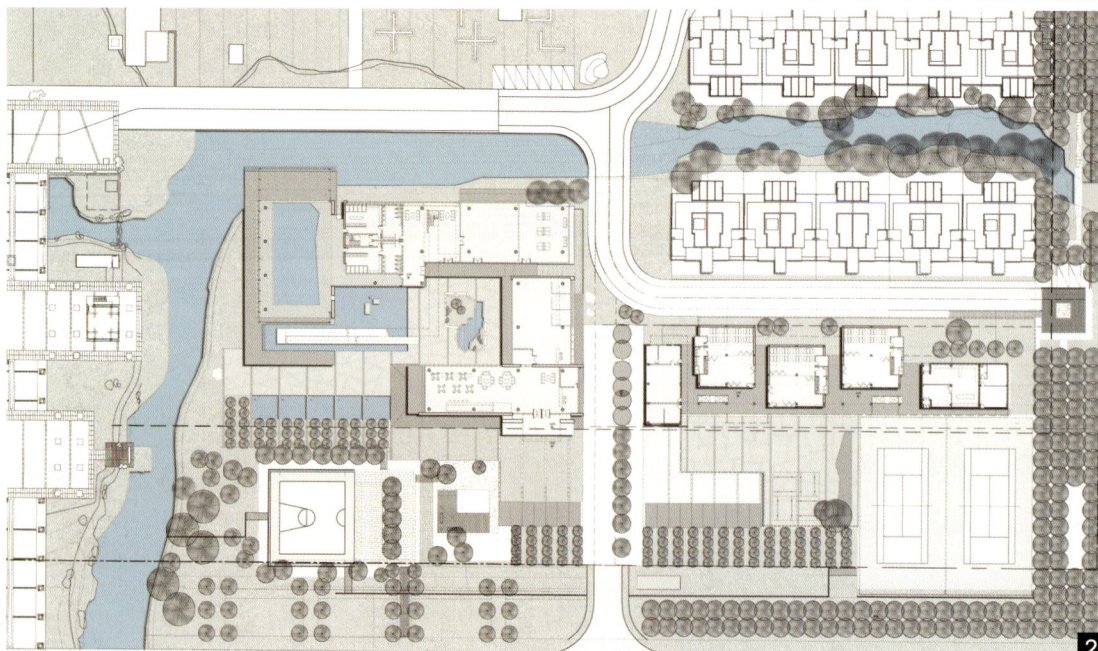

2

Evian Town Phase II
依云水岸二期

Photography: Hans-Georg Esch

1. Clubhouse Internal Courtyard
   会所内园

2. Plan
   平面图

3. Clubhouse Internal Courtyard
   会所内园

4. Classroom
   校园课堂

Evian Town Phase I
依云水岸一期

Photography: But-Sou Lai

1. Retail Internal Courtyard
   商业内园

2. Clubhouse Internal Courtyard
   会所内园

3. Clubhouse Internal Courtyard
   会所内园

3

Linked Hybrid

北京当代万国城

The 220,000-square-meter Linked Hybrid complex in Beijing, aims to counter the current privatized urban developments in China by creating a 21st century porous urban space, inviting and open to the public from every side.

The ground level offers a number of open passages for all people (residents and visitors) to walk through. These passages include "micro-urbanisms" of small-scale shops which also activate the urban space, surrounding the large central reflecting pond. On the intermediate level of the lower buildings, public roof gardens offer tranquil green spaces, and at the top of the eight residential towers private roof gardens are connected to the penthouses. All public functions on the ground level, – including a restaurant, hotel, Montessori school, kindergarten, and cinema – have connections with the green spaces surrounding and penetrating the project.

Geo-thermal wells (655 at 100 meters deep) provide Linked Hybrid with cooling in summer and heating in winter, and make Linked Hybrid one of the largest green residential projects. The large urban space in the center of the project is activated by a greywater recycling pond with water lilies and grasses in which the cinematheque and the hotel appear to float. In the winter the pool freezes to become an ice-skating rink.

The water in the whole project is recycled. This greywater is piped into tanks with ultraviolet filters, and then put back into the large reflecting pond and used to water the landscapes. Re-using the earth excavated from the new construction, five landscaped mounds to the north contain recreational functions. The "Mound of Childhood", integrated with the kindergarten, has an entrance portal through it. The "Mound of Adolescence" holds a basketball court, a roller blade and skate board area.

22万平方米的北京当代万国城，旨在通过创造21世纪多元都市模式来改变现有的私有化城市风格，并将其从各个方面对公众开发和被接纳。

项目的一层设计了一套步行系统，一个既为社区居民又为社区外民众提供的游乐之所。这些场所包含一些围绕在中心水池四周的活跃的小商铺所形成的"城市综合体"。在大厦的中下部分，开放式的屋顶花园提供了绿色空间，并且八个住宅塔楼的私人花园还被连接到了大厦顶端。底层的所有公共功能区——包括餐厅、宾馆、学校、幼稚园和电影院等——都与周围贯穿整个项目的空中连桥相连接。

665个100米深的地热井使北京当代万国城冬暖夏凉，并且成为最大的绿色环保住宅项目。项目中心的巨大中水回收池，种植着荷花和水草，摇曳之时看去仿佛电影院和酒店等都像在水上漂浮，而冬季这里将凝结成一个巨大的溜冰场。

整个项目的水都是可回收的，这些中水经过管道被回收并在系外线过滤之后重新回到中心的水池作为景观装点之用。同样，利用新工程挖掘出的泥土，在园区北侧形成了五个小山，美化了环境的同时也增添了其他的功能性。

Project name: Linked Hybrid
Award date: 2010
Location: Beijing, China
Floor area: 221,462 m²
Architect: Steven Holl Architects, Beijing Capital Engineering Architecture Design Co., Ltd.
Client: Modern Green Development (Beijing) Co., Ltd.
Photographer: Steven Holl Architects, Iwan Baan, Shu He
Completion date: 2009
Award name: McGraw-Hill Construction 3rd Bi-Annual "Good Design Is Good Business" China Awards 2010, Best Residential Project

项目名称：北京当代万国城
获奖时间：2010
项目位置：中国 北京
项目面积：221,462平方米
建筑设计：Steven Holl Architects   北京首都工程建筑设计有限公司
业主：北京当代集团
摄影师：Steven Holl Architects Iwan Baan   舒赫
完成时间：2009
所获奖项：麦格劳–希尔公司《建筑实录》、《商业周刊》第三届"好设计创造好效益"中国奖项   2010最佳住宅建筑

1

2

3

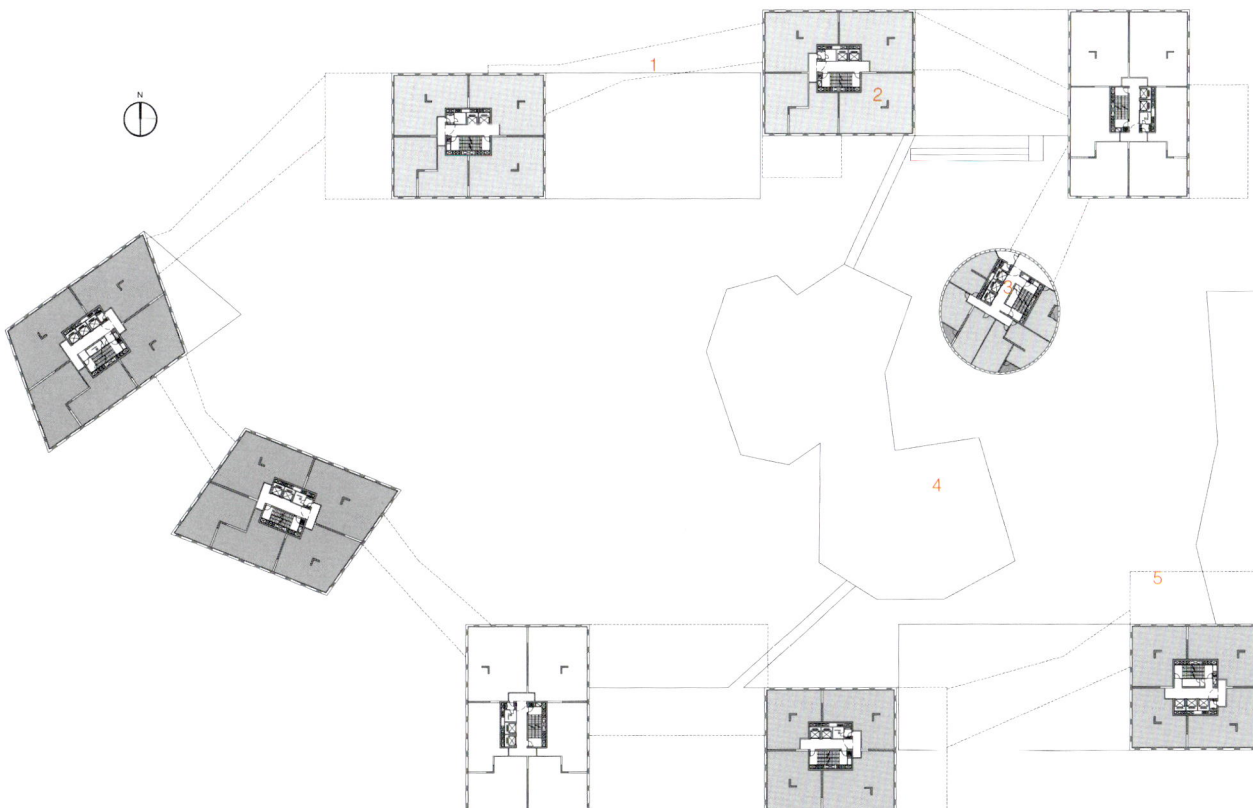

1. Bridges
   空中连桥
2. Towers
   高层建筑
3. Hotel
   宾馆
4. Cinema with Public
   Roof Garden
   电影院及公共屋顶花园
5. Restaurant
   餐厅

4

1. Full View
   全景
2. Plan
   平面图
3. Up View of the Linked Bridge
   天街步道仰视图
4. Detail
   大厦一角

1. Interior of the Bridge
天桥内部

2. Interior Stairs
室内台阶

3. Plans
平面图

4 Theater
剧场

3

4

Ningbo Eastern New City Economical Housing

宁波东部新城安置住宅区

The temporary housing for demolition is located in the northwest corner in the center of the eastern new town. The site, with an area of 19.77 hectares, is situated in the south of Qinzhou High School. The total floor area above ground, including housing and other facilities such as a kindergarten (with twelve classes), clubs, a supermarket and a market, is 300,000 square meters, while the total floor area of underground and semi-underground parking is 60,000 square meters. About 11,000 people are scheduled to reside there.

These housing units are designed with respect to the living habits of their future occupants; meanwhile, new living concepts such as health, ecology and sport are introduced. Besides, advanced technologies and materials, for example, EPS polyphenyl thermal panels, insulating glass, integrated service management network and far-infrared alarm device are adopted.

For each building, various design elements are integrated, including opening, abstract painting, sky garden, color and frame. Large terraces made vertical landscaping possible, changing the landscape of the housing complex from two-dimension to three-dimension. In the higher levels, terraces appear as orderly patterns on the facades, offering distinctive visual experience.

The conception and construction of the project in the new eastern town is a successful attempt in many aspects. The relocation of the residents has been proved socially and economically beneficial. The architects visited five households, and they expressed their satisfaction and said that this was the best temporary housing they'd ever seen. Moreover, the housing has brought them tangible benefit: in the second-hand housing market, the housing rate has risen from 6,800 RMB/m² to 15,000 RMB/m².

东部新城拆迁安置房位于东部新城核心区东北角，鄞州中学以南，占地面积19.77公顷，地上住宅及配套设施(包括12个班的幼儿园,商业会所,超市,净菜市场等)建筑面积30万平方米，地下和半地下停车库6万平方米，计划安置拆迁居民11000人。

东部新城拆迁安置房充分关注安置对象原有居住习惯和社会细节的变化，引入"健康、生态、运动"等新理念和"居住岛"设计概念，并应用先进的EPS聚苯保温板、中空玻璃等新型科技材料，设置了网络综合管理服务系统、远红外线防盗报警装置。

在单体设计当中，可以提取的设计元素是：开洞，抽象画，空中花园，色彩，廊架。建筑大面积的露台提供了垂直绿化的可能，使小区整体的景观体系由二维走向三维。露台在小高层的建筑体量上以构图元素的形式出现，错落有秩，形成不同一般的视觉体验。

东部新城安置住宅的设计和建设是一次成功的尝试，在目前正在进行的安置过程中，收到了良好的社会效益和经济效益。在建成后的建筑师回访过程中，受访的五个安置村庄农民普遍表示感到满意，承认这是他们见过的最好的安置住宅，并已经给他们带来了实惠，二手房市场的房价已经从刚交房时的6800元每平方米上涨到15000元每平方米。

Project name: Ningbo Eastern New City Economical Housing
Award date: 2008
Location: Ningbo, China
Building area: 300,000 m²
Architect: DC ALLIANCE, China Ningbo Housing Design Institute
Client: Ningbo Eastern New City Development & Construction Headquarters
Completion date: 2007
Award name: McGraw-Hill Construction 2nd Bi-Annual "Good Design Is Good Business" China Awards 2008, Best Residential Project

项目名称：宁波东部新城安置住宅区
获奖时间：2008
项目位置：中国 宁波
建筑面积：300,000平方米
建筑设计：DC国际 宁波房屋建筑设计院
业主：宁波东部新城开发建设指挥部
完成时间：2007
所获奖项：麦格劳-希尔《建筑实录》、《商业周刊》第二届"好设计创造好效益"中国奖项   2008最佳住宅建筑

1. Green and Plants
   绿地及植被

2. Public Space
   公共空间

3. Pathway
   人行道

4. Section
   剖面图

1

2

1. Main Gate
   主入口

2. Plan
   平面图

3. View from Water
   亲水楼盘

Southside House in Hong Kong

香港南海岸独栋住宅

Chang Bene's client, a Hong Kong businessman with a penchant for the outdoors and swimming, wasn't interested in increasing the size of his 3,500-square-foot residence, but demanded a space to hold informal meetings and entertain clients with an outdoor pool and patio. Due to the limited space, "It was like a puzzle game," says Shirley Chang, one of the two principals at Chang Bene. "All the possible designs had trade offs. There was constant bargaining." The original building was a two story home that sat above a carport and a small ground floor. In addition, each floor was divided into small, enclosed rooms, a typical arrangement in Hong Kong homes, many of which are limited in square footage. Chang Bene enclosed the existing carport and broke through its ceiling to create a two-story high living room. A large, retractable glass shutter is placed between the living room and the new lap pool outside to make the living room into a "floating pavilion". The architects designed the ground floor to work as one large entertaining space, with a living room, open kitchen, and enclosed dining deck, all flowing into each other. Above the ground level, Chang Bene created a library mezzanine that overlooks the living room. On the top level, the architects converted three separate bedrooms into a single master bedroom and bathroom with lacquered floor-to-ceiling panels that slide open to increase cross ventilation and light. "The hardest part of the project," says Chang, "was convincing the Building Department of Hong Kong to approve the two-story-high living room." Because of the limited space in the city, the department was "suspicious" of such an idea, in disbelief that a resident would prefer a high-ceilinged living room to more square footage. "The two-year project was also atypical for Hong Kong," says Chang, "because the client allowed the firm to take as much time as it needed." Christopher Bene, Chang's partner, says, "Increasingly, people don't want chopped rooms, they want light and air. They want to connect with the ground plane." Chang adds, "The lesson is that [some people] value good design."

Chang Bene设计事务所的客户是一位香港商人，热衷户外活动和游泳，要求在占地325平方米的住宅建筑内增设一个室外游泳池和露台，以供招待客户之用。事务所的负责人之一Shirley Chang曾这样说道："在这样一个有限的空间内进行设计仿佛在做一个益智游戏。"项目在经过多方周密研究之后才得以顺利进行。原建筑共两层，负一层设有停车库和小型地下室。另外，每个楼层均设有小型封闭较好的房间，是很典型的香港住宅布局。设计师将原有车库封闭之后，将地下室与一楼间的天花板打通，构建了一个两层的客厅。在客厅与户外泳池中间打造一个大型可伸缩玻璃窗，使客厅犹如一个"浮动亭"。地下一层主要作为娱乐空间，除客厅之外，还设有开放式厨房、封闭式饭厅，各区间衔接自然、和谐相通。从一楼的图书馆中层楼能够俯瞰整个客厅。楼上原有3个独立的卧室打通之后成为一个主卧室，其中设有浴室，漆面落地式滑动镶板便于室内的通风和采光。该项目最具挑战的是说服香港营建署对于两层高客厅的审批。在这个寸土寸金的城市中，住户在不扩充建筑面积的前提下将客厅的天花板调高着实令营建署质疑。客户并未对项目的竣工时间加以限定，该工程历时两年，如此长的周期在香港的建筑设计中并不多见。Chang Bene设计事务所的另一位负责人Christopher Bene说："越来越多的客户想要打破小房间的格局，更向往通透，采光和通风性能良好的空间，与地面层衔接自然。"Shirley Chang补充道："人们对美的东西总是难以抗拒的。"

Project name: Southside House in Hong Kong
Award date: 2008
Location: Hong Kong, China
Building area: 320 m²
Architect: Chang Bene Design
Client: C. Tse
Photographer: Chang Bene Design Ltd
Completion date: 2005
Award name: McGraw-Hill Construction 2nd Bi-Annual "Good Design Is Good Business" China Awards 2008, Best Residential Project

项目名称：香港南海岸独栋住宅
获奖时间：2008
项目位置：中国 香港
建筑面积：320平方米
建筑设计：Chang Bene设计事务所
业主：C. Tse
摄影师：张贝理设计事务所
完成时间：2005
所获奖项：麦格劳-希尔公司《建筑实录》、《商业周刊》第二届"好设计创造好效益"中国奖项　2008最佳住宅建筑

1. Living Room
   起居室
2. Sun Room
   阳光房

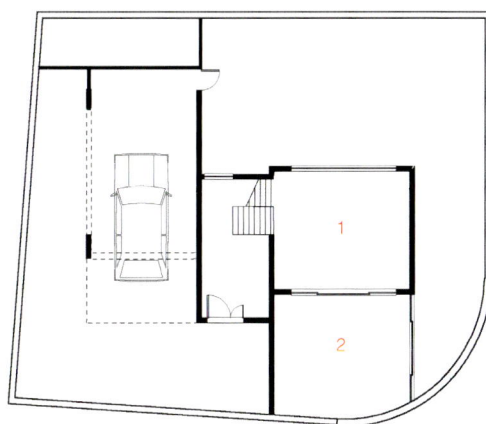

1. Dining Deck
   餐厅

2. Living Room
   起居室

3. Living Room
   起居室

4. Plans
   平面图

4

1. Living Room
起居室

2. Corner Window
角窗

3. Study Stairs
楼梯

4. Kitchen and Dining
厨房及餐厅

5. Kitchen
厨房

The Dutch Ambassador's Residence

荷兰大使官邸

The Ambassador's residence in Beijing houses three functions: representational space, a service area and private space. Because of the relatively large site (4,220 m²), it was possible to create a single story building with direct access to each function without jeopardizing the privacy of the living quarters. The building has two wings: one contains the meeting and dining rooms and the service area; the other houses the private quarters. Each wing has its own orientation towards the garden. Both wings are connected by a winter garden. This space offers a permanent representation of nature (and proof of the Dutch ability to keep that nature under a glass roof), shielded from the extremes of the Beijing climate. A large entry gives access to the representative part of the building, with a direct view through the building into the garden. Ample room is given to the ritual of welcoming guests, taking coats, presenting the guestbook, etc. From the entrance hall, a large reception room facing east, a small and intimate "salon" and a large dining room can be reached. The large cantilevered roof protects the interior from direct sunshine in the hot season, and provides a dry terrace in case of rainfall. In the private wing the kitchen and the main room are the central spaces in a linear plan. The three bedrooms are situated at the far end for maximum privacy. On special occasions, the ambassador can take a guest through the winter garden to his study/library in the private wing. The main expression of the house is dominated by an elongated wall that stretches beyond the building itself. The wall is made of horizontal strips of natural stone of different dimensions, and has strips of lighting integrated into it. The entry is shielded by a large pane of safety glass. The garden elevation consists of structural glazing and sliding doors. The private wing is enclosed by a stone wall. On the west side the rooms open up towards the private garden. Louvers provide sun-shielding and privacy. The horizontal articulation enhances the sense of spaciousness, and counters the verticality of the trees.

Since the opening in 2007, a new ambassador has moved in. The family prefers to use the representational rooms equally for their daily life. Official gatherings have proven to be more popular than envisaged. On Queensday (a Dutch national holiday) over 500 people enjoy the hospitality of the residence. However, the house facilitates this very well.

由Kraaijvanger Urbis设计事务所设计的荷兰大使官邸坐落于4,220平方米的开阔场地上，集会客、服务、住宅三种功能于一体。该建筑不但为大使及其家人提供居住空间，同时对伟大的荷兰国度也具有一定的象征意义。大气、奢华的独立式住宅十分引人注目。这座建筑有两座附楼：一座是会客室、餐厅和服务区，一座是私人居住空间。每一座附楼均面向花园，并经由花园自然衔接在一起。考虑到北京严寒的冬季，设计师巧妙地将这个花园设计成透明的室内温室结构。开阔的室内空间专为会客之用。入口大厅处是一个东向的接待室，与小型私密"沙龙"以及大型餐厅相通。宽大的悬垂式屋顶能够完美地遮挡烈日对室内的照射，同时可以在雨季适当排水。私人居住空间中，厨房和主室位于中央，成线性布局。三间卧室位于空间的一端。在特殊场合，大使可以带领客人穿过花园直达私人居住空间中的书房。造型别致的外墙有效遮挡了内部空间，并朝向花园打开。入口处设有大型安全玻璃，时刻确保室内的通透。花园的正面设有结构性玻璃和推拉门。私人住宅空间外设有围墙。西侧的房间与私人花园相通。精致的百叶窗能够有效遮挡室外光线，并营造私密氛围。水平方向的延伸提高了空间感，同时和树木的垂直形成了对比。

该建筑自2007年竣工以来，已有一名大使及其家人入住，起初他们只是将其作为日常生活空间，而实际上这里已成为官方聚会的完美场所。在荷兰女王日（荷兰国定假日）的当天，这里将为500多人提供热情款待，该建筑的结构设计也为宴会的举办提供了优越条件。

Project name: The Dutch Ambassador's Residence
Awarded date: 2008
Location: Beijing, China
Gross floor area: 850 m²
Architect: Dirk Jan Postel/Kraaijvanger-Urbis, Universal Architecture Studio and Royal Haskoning
Client: The Kingdom of The Netherlands
Photographer: Christian Richters
Completion date: 2007
Award name: McGraw-Hill Construction 2nd Bi-Annual "Good Design Is Good Business" China Awards 2008, Best Residential Project

项目名称：荷兰大使官邸
获奖时间：2008年
项目位置：中国 北京
总面积：850平方米
建筑设计：Kraaijvanger Urbis设计事务所
业主：荷兰王国
摄影师：Christian Richters
完成时间：2007年
奖项名称：麦格劳-希尔公司《建筑实录》、《商业周刊》第二届"好设计创造好效益"中国奖项　2008最佳住宅建筑

1. Niight View
   夜景
2. Cantilevered Roof over the Elevated Terrace
   一层露台上的悬顶
3. Granite Wall, Timber Roof and Glass Gacade
   石墙、木顶及玻璃正面
4. Plan
   平面图

1. Representative Entrance
   代表入口
2. Dressing Room
   衣帽间
3. Reception Room
   接待室
4. Small Reception Room
   小接待室
5. Dinning Room
   餐厅
6. Kitchen
   厨房
7. Terrace
   平台
8. Winter Garden
   冬日花园
9. Private Entrance
   私人入口
10. Garage
    车库
11. Bedroom
    卧室
12. Library
    资料室
13. Living Room With Open Kitchen
    起居室和开放式厨房

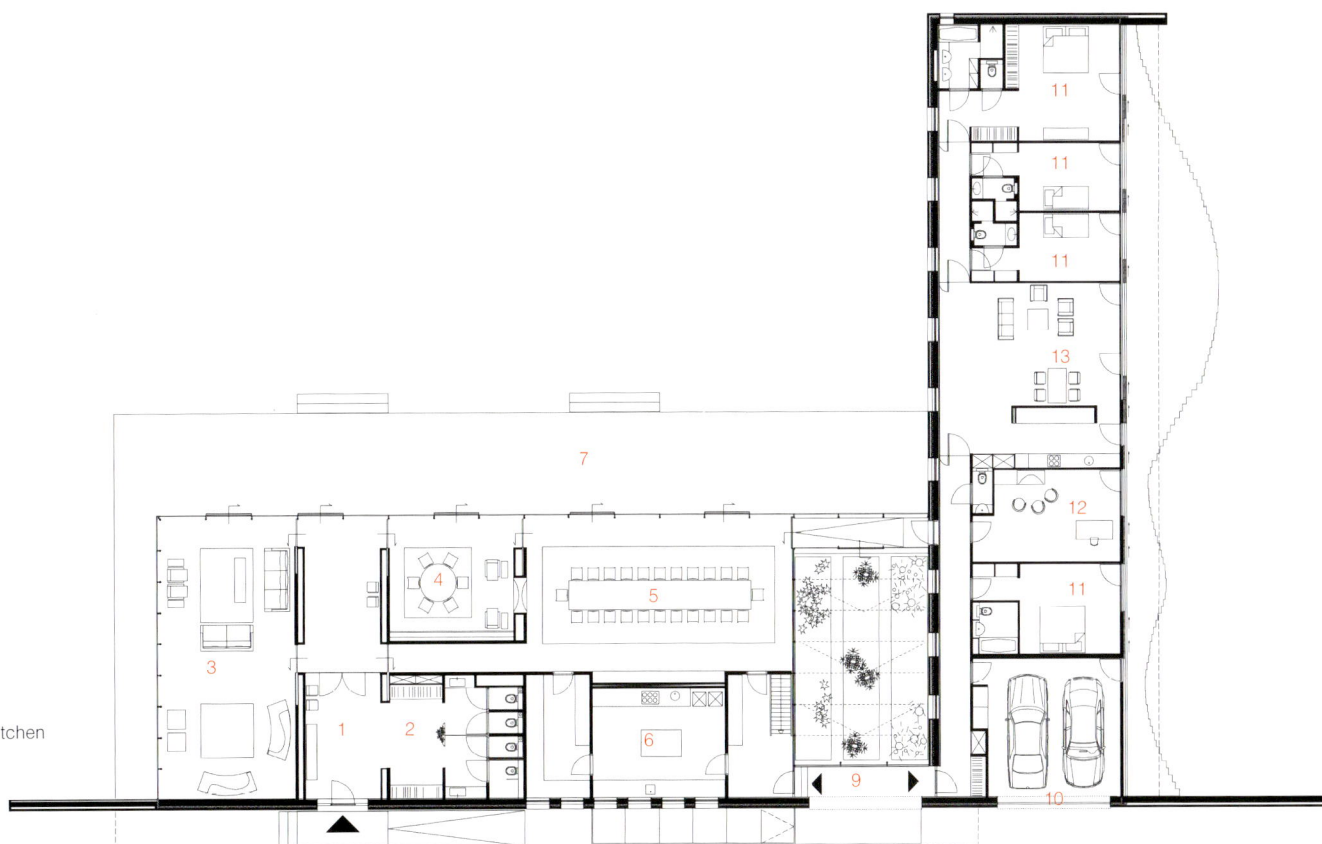

1. Blending of Interior and Exterior Spaces
   室内与室外空间的融合

2. Exterior View of Private Wing + Garden
   私人区及花园外景

3. Main Wall and Roof
   主外墙及悬顶

4. Night View
   夜景

5. Front Court with Visually Open Fence
   开放式围栏的庭院

1. Granite Wall Extended into the Interior
   石墙连通室内

2. Large Reception Room at the East
   东侧的大接待室

3. Full Depth of the House
   房间纵深透视

4. Main Dining Room and Tropical Winter Garden
   主餐厅及花园

Jianwai SOHO

建外SOHO

Jianwai SOHO is situated in the Central Business District (CBD) of Beijing, two kilometers from Tian'an Men Square in the city center. The complex comprises apartments, offices and shops, with a gross floor area of more than 700,000 square meters. On the site of 169,000 square meters, the slim towers, all square in plan (with a length of 27.3 meters), twist 25 degrees from the south-north direction in order to enjoy better natural lighting and to avoid direct view between every two buildings. The tallest tower reaches a height of 100 meters. Driveways and sidewalks are separated, and the sunk plazas among the towers respond to the surrounding landscape. These carefully-considered details made possible the creation of a unique landmark in Beijing, contributing to a more diverse street scenery.

Each unit is a single SOHO. Apart from offices, they serve with all other facilities except housing. Public spaces are not for occupants only; all citizens could enjoy their leisure there. Jianwai SOHO has been highly praised for its simple design, detailedly-planned connections among single towers, and its contribution for a new way of urban life. As a symbol of "Fashion Beijing", Jianwai SOHO has attracted wide attention from all walks of life. When construction has been finished for six years, it still appears frequently through various kinds of media as the favorite meeting place of Beijing citizens.

建外SOHO距离北京市中心的天安门广场有2公里，位于被称为CBD的商贸中心区，由住宅、办公、店铺构成综合建筑群规划。总建筑面积超过70万平方米的地块中，林立着27.3米见方、最高处达100米的细高塔楼。考虑到采光均匀及避免楼栋间对视问题，每栋塔楼都沿正南北方向偏转了25°；车辆和步行者交通流线完全分开；自由穿插在地块内的下沉广场和地面的园林景观相呼应；这些设计细节处理造就了北京独一无二的风景，呈现丰富的街道风貌。

每个单元都是SOHO，不仅可以办公，还具备了除住宅功能之外的其他设施。公共部分不仅为在此居住的人专设，它是提供给市民自由使用的广场。建外SOHO通过它纯净的设计、精细的节点处理，提供了新生活方式构想等，获得了很高的评价。作为"时尚北京"的代表地，被各方人士关注。在竣工6年之后，仍然频繁在各种媒体上登场，是北京市民喜爱的聚集地。

Project name: Jianwai SOHO
Award date: 2006
Location: Beijing, China
Building area: 504,237 m²
Architect: Riken Yamamoto & Field Shop, C+A, and Mikan
Client: SOHO China Ltd
Completion date: 2003
Award name: McGraw-Hill Construction 1st Bi-Annual "Good Design Is Good Business" China Awards 2006, Best Residential Project

项目名称：建外SOHO
获奖时间：2006
项目位置：中国 北京
建筑面积：504,237平方米
建筑设计：山本理显设计工场　C+A　Mikan
业主：SOHO中国有限公司
完成时间：2003
所获奖项：麦格劳-希尔公司《建筑实录》、《商业周刊》第一届"好设计创造好效益"中国奖项　2006最佳住宅建筑

1

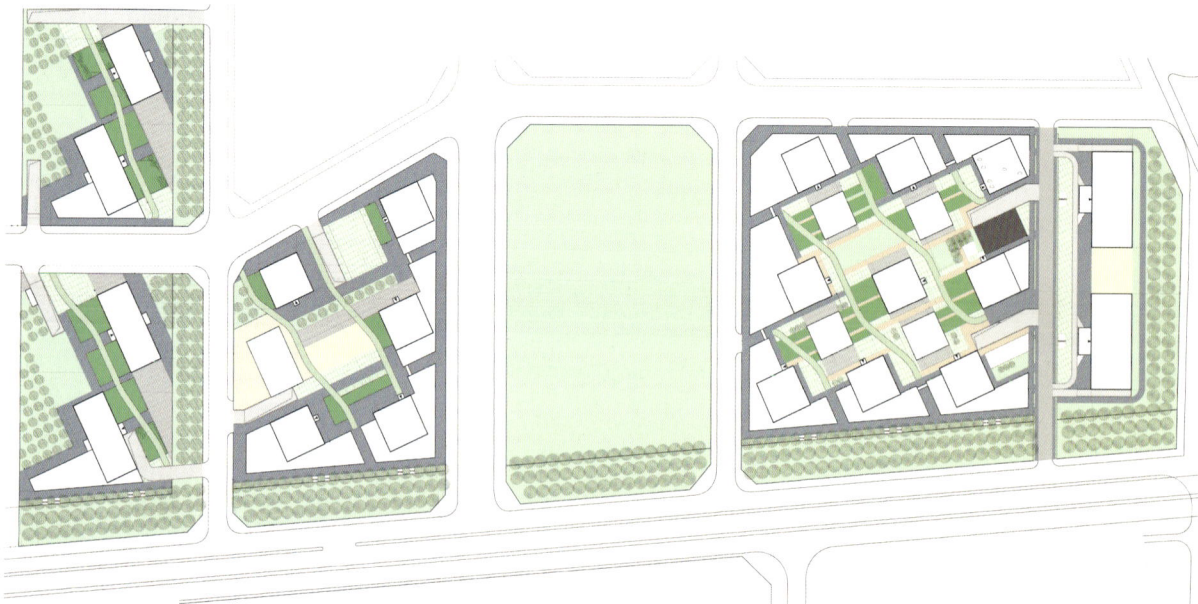

2

1. View from the River
   远景

2. Plan
   平面图

3. Living Room
   室内起居室

4. Interior
   室内

5. Interior
   室内

Villa Shizilin

柿子林会馆

The client is a developer couple with two children and they have acquired a piece of land in the adjacency of the Ming Tombs outside Beijing to build a house with extensive programs from cinema to indoor swimming pool, with the intention to use the facilities also as a club.

The building site used to be an orchard for persimmon trees surrounded by mountains. To fully engage the sceneries in the area, the architecture design takes the rangefinder as concept and nine tapered spaces are oriented towards nine different views. The roofs slope to complete these perspectival rooms while to interpret the traditional Chinese architectural forms in a topological way. By the same token, a rolling artificial landscape is created on the rooftop which echoes the hilly terrain nearby. The structure of the building is a concrete sheer wall and beam system with local granite clad on the outside of the walls and dark cement tiles on the roofs. The existing persimmon trees are preserved to punctuate the house.

经过了四轮的设计，这个在北京昌平一个柿子林中的项目终于聚焦在看与被看这一对关系上：由于建筑周围有优美的自然环境，于是房间或房间组被作为取景器来设计。取景器——房间共有九个，面向不同的方向与景观；看的需要促使其三维形状内收外放，作为景框的大口是落地窗，两侧承重实墙呈八字关系，屋顶倾斜构成单坡。换而言之，坡屋面首先是为限定取景器而出现的。取景器–房间之间，有时是中间，则是保留下来的柿子树。建筑与景观又相互融合了，建立起与基地之间的另一重关系。

其实，建筑同时又总是被看的对象。建筑设计也是造景。将九个取景器——房间倾斜程度不同的屋顶整体看待，建筑顶部便出现了一个起伏的相对复杂的拓扑接口；也许可以认为是一个人造的地景，与基地周围的山峦呼应；也许又可以作为以当代的建筑语言翻译传统中国建筑坡顶形式的一次尝试。以往中国建筑的研究中，更偏重于单体屋顶形式的传承。而典型的中国建筑则普遍是以群体存在的。一个院落四周建筑的屋顶大小变化本身便包含了拓扑关系。在柿子林，取景器是决定屋顶形式的先决，而对中国传统建筑群体屋顶的观察提供了最终的参考。

柿子林别墅/会馆采用与取景器空间完全吻合的、不平行石夹混凝土承重墙与混凝土反梁结合的结构体系。因此结构是建筑不可分割的一部分。石料为就地取材的花岗岩。长期以来，鉴于前辈建筑师在中国建筑的形式与形象方面做了大量的深入的工作，非常建筑则将对传统的探索转向空间和建造等问题，对形式与形象一直没有形成切入点。群体屋顶拓扑目前是我们形式工作的突破口，向建立中国建筑的当代性与地域性全面推进。

Project name: Villa Shizilin
Award date: 2006
Location: Beijing, China
Building area: 4,800 m²
Architect: Atelier Feichang Jianzhu
Designer: Zhang Yonghe, Wang Hui
Client: Antaeus Group
Photographer: Shu He, Fu Xing
Completion date: 2004
Award name: McGraw-Hill Construction 1st Bi-Annual "Good Design Is Good Business" China Awards 2006, Best Residential Project

项目名称：柿子林会馆
获奖时间：2006
项目位置：中国 北京
建筑面积：4,800平方米
建筑设计：非常建筑
设计主持人：张永和　王晖
业主：今典集团
摄影师：舒赫　付兴
完成时间：2004
所获奖项：麦格劳–希尔公司《建筑实录》、《商业周刊》第一届"好设计创造好效益"中国奖项　2006最佳住宅建筑

1. East View
   东侧全景

2. Main Entrance
   主入口

3. East Corner form South
   南侧东端

4. Outside of Swimming Pool
   游泳馆外幕墙

1. First Floor Corridor
   一层过厅

2. First Floor Dining Hall
   一层餐厅

3. Second Floor Terrace
   二层楼顶平台

4. Second Floor Interior
   二层内部

5. Plan
   平面图

1. Servant Bedroom
   佣人卧房
2. Guest Bedroom
   客房
3. Guest Suite
   客人套房
4. Sun Deck
   起居阳台
5. Glass Roof
   玻璃顶
6. Children's Living Room
   儿童起居室
7. Children's Bedroom
   儿童卧室
8. Study Room
   书房
9. Nanny Room
   保姆房
10. Master Sitting Room
    主起居室
11. Master Bedroom
    主卧室
12. Storage
    储藏室
13. Balcony
    阳台
14. Toilet
    卫生间
15. Bathroom
    室外洗浴

Longyang Residential Complex

东方汇景苑

The pressures of the large-scale housing market conspire to create apartment housing that is often repetitive and banal. In the place of clever design, advertisements – which often invoke the kitch and the "classical" rather than the innovative and modern – become the vehicles for selling-off apartments. The Longyang development attempts to subvert this traditional approach to apartment housing design by fulfilling a bi-fold mandate:

1) to break-up the normative homogeneity both in and outside the apartments and
2) to maximize the amount of light and green space in the building area.

The former was achieved through alternating patterns, shapes and coloration on the building facade. The balconies on each level alternate between an enclosed shaft that offers a framed view and an open deck which offers views through a clean, large projection of space. Duplexes nest in the top floors of every other building, and their elevation differs significantly from those of the smaller apartment facades. The buildings' plans also break the monotony by literally bending the traditional straight-bar plan into alternating building "cups". When paired with the building directly before or after it, the negative space forms large communal green areas that either seem enclosed by the buildings or opened-up to the street. By situating the taller structures to the North of the development, lengthy shadows fall into the street rather than the adjacent building.

The combined effect of these subtle tweaks to the normative building pattern give the development's inhabitants the feeling that they are much more than cogs in a monochromatic, mono-textural machine, but rather owners of a unique piece of urban real estate.

大量成片开发的住宅的设计在市场的压力下，往往变成一件重复且平庸的工作。聪明的开发商，经常采用促销广告的手段来制造产品所谓的"经典"或"创新"模式，而不去更多的关注居住状态的创新和建筑物的现代化。东方汇景苑利用以下两点来尝试去消弱这个住宅设计的习惯。

1）从内部和外部两方面来打破标准化的同质性；
2）在住宅楼中实现绿化和采光的最大化。

前者通过建筑中重复的元素交替的位置、形状和外立面的不同色质来实现。譬如每一层的阳台左右的位置变化，及上下不同的围合方式，时而形成景观视窗，时而形成宽敞无阻的视野。每隔一栋住宅的顶层设有一跃式层，其立面也力图打破整体立面的统一感。与其他小面积房型的立面有着显著的不同。在建筑平面上也在日光允许的范围之内变化角度，从而打破了传统板楼的单一平面形式，也改变了开放空间的格局。这些是将住宅楼之间的公共空间，不断地变换着它的围合感，同时也加强了住户的认同感。这个系列的开敞空间继而再融入更高一级的社区绿地空间。这些空间面向南，享有充分的阳光。

这一切微妙的思考所产生的效果，相对于其他标准化的住宅模式，带给住户更多独特都市房地产商品的舒适、人性化感受，而不是仅仅拥挤于无质感的单元机械重复的单调乏味。

Project name: Longyang Residential Complex
Award date: 2006
Location: Shanghai, China
Site area: 52,000 m²
Architect: MADA s.p.a.m.
Client: Shanghai Kangwei Development Co. Ltd.
Photographer: Jin Zhan
Comepletion date: 2003
Award name: McGraw-Hill Construction 1st Bi-Annual "Good Design Is Good Business" China Awards 2006, Best Residential Project

项目名称：东方汇景苑
获奖时间：2006
项目位置：中国 上海
占地面积：52，000平方米
建筑设计：马达思班建筑设计事务所
业主：上海康伟置业有限公司
摄影师：金雷
完成时间：2003
所获奖项：麦格劳-希尔公司《建筑实录》、《商业周刊》第一届"好设计创造好效益"中国奖项　2006最佳住宅建筑

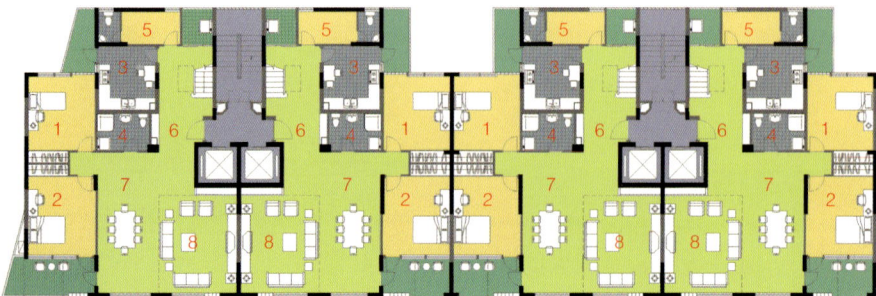

1. Bedroom
   卧室
2. Master Bedroom
   主卧室
3. Kitchen
   厨房
4. Washroom
   卫生间
5. Study Room
   工作间
6. Entrance Hall
   门厅
7. Dining
   餐厅
8. Living Room
   起居室

1. Facade of the Apartment
   正面图

2. Plans
   平面图

3. Part of the Apartment
   楼盘一角

4. Part of the Apartment
   楼盘一角

Oriental Garden

东方庭院

These villas and townhouse are designed for those families or individuals who would like to live a modern lifestyle as well as being in a close relationship with nature, and here the relationship between people and water is emphasized.

The historic Zhujiajiao old Water Town is right next to the site. It is one of the four famous old cultural towns of Shanghai, and has come to be know as the Venice of Shanghai. Three sides of the site are surrounded by water, and a river goes through the site from the northwest corner.

To reflect and emphasize the character of a water town, new canals are built along/around the old ones. Thus a center of the village is formed with townhouses sitting around by the water. Water Alleys and Street Alleys are the fabrics of a traditional water town in southeast China. Water Alleys performed in the past as the main channels for traffic, commodity transportation and the connections to the neighbor village. Stores on both sides of the Street Alleys were inseparably next to each other. All these Street Alleys were for people to shop, to communicate, and to socialize. The pedestrian area seems narrow and irregular.

这些别墅和联排房屋是为了那些喜欢现代生活风格又钟爱自然的家庭或个人而设计的，我们强调的是人与水之间的和谐关系。

历史悠久的朱家角水乡在本案旁边。它是上海四大著名的文化古镇之一，被誉为上海威尼斯。这里三面环水，一条河流从西北角流入此地。

为了表现和强调水乡的特色，新的运河被修建在旧运河之上或者周围。因而城镇的中心也坐落于水上的联排房屋。在中国江南水乡，水路与沙石街道的胡同是典型传统的风格。水路在过去充当着交通、运输以及与临镇相连接的要道，街道两旁的店面也都是紧邻彼此不可分割的。所有这些街道胡同都是为了人们购物、沟通及社交需要而建的。作为人行道的部分，则相对狭窄和不规则。

Project name: Oriental Garden
Award date: 2006
Location: Shanghai, China
Building area: 227, 214 m²
Architect: Benwood Studio Shanghai/Benjamin Wood
Client: SPG Land (Holdings) Limited
Completion date: 2005
Award name: McGraw-Hill Construction 1st Bi-Annual "Good Design Is Good Business" China Awards 2006, Best Residential Project

项目名称：东方庭院
获奖时间：2006
项目位置：中国上海
建筑面积：227，214平方米
建筑设计：Ben Wood 上海工作室
业主：盛高置地（控股）有限公司
完成时间：2005
所获奖项：麦格劳–希尔公司《建筑实录》、《商业周刊》第一届"好设计创造好效益"中国奖项 2006最佳住宅建筑

1

1. Sales
   售楼中心
2. Villa
   会所
3. Double House
   叠拼别墅
4. United House
   联排别墅
5. Single House
   独栋别墅

2

1. Sales Exterior
   会所
2. Master Plan
   总平面图
3. Villa Plans
   楼盘图纸
4. Exterior
   别墅外景
5. Exterior
   别墅外景

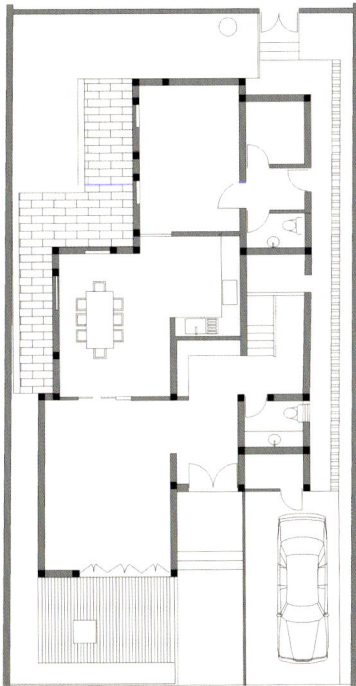

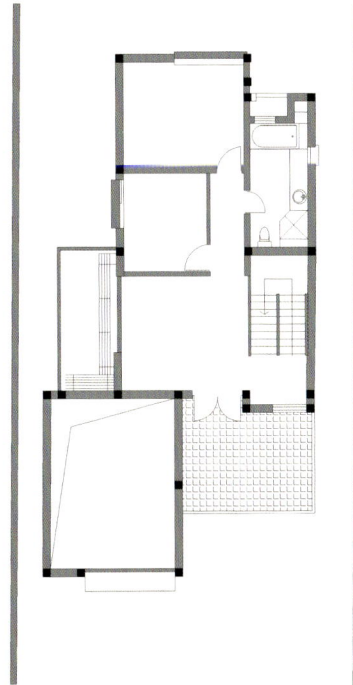

3

4

5

1. Full View
全景

2. Path near Water
水边小路

3. Steps near Water
水边台阶

4. Bird View
鸟瞰

1. Water
   水景

2. Exterior
   外景

3. Interior Landscape
   会所院内景观

4. Exterior
   外景

5. Sales Interior
   会所室内

Hutong Bubble 32

胡同泡泡32号

Beijing 2050 imagined three scenarios for the future of Beijing – a green public park in Tian'an Men Square, a series of floating islands above the city's CBD, and the "Future of Hutongs", which featured metallic bubbles scattered over Beijing's oldest neighborhoods. Three years later, the first hutong bubble has appeared in a small courtyard in Beijing.

China's rapid development has altered the city's landscape on a massive scale, continually eroding the delicate urban tissue of old Beijing. Such dramatic changes have forced an aging architecture to rely on chaotic, spontaneous renovations to survive the ever-changing neighborhood.

The self-perpetuating degradation of the city's urban tissue requires a change in the living conditions of local residents. Progress does not necessarily call for large-scale construction – it can occur as interventions at a small scale. The hutong bubbles, inserted into the urban fabric, function like magnets, attracting new people, activities, and resources to reactivate the entire neighborhoods. They exist in symbiosis with the old housing. Fueled by the energy they helped to renew, the bubbles multiply and morph to provide for the community's various needs, thereby allowing local residents to continue living in these old neighborhoods. In time, these interventions will become part of Beijing's long history and newly formed membranes within the city's urban tissue.

Hutong Bubble 32 provides a toilet and a staircase that extends onto a roof terrace for a newly renovated courtyard house. Its shiny exterior renders it an alien creature, and yet at the same time, reflects the surrounding wood, brick, and greenery. The past and the future can thus coexist in a finite, yet dream-like world.

The real dream, however, is for the hutong bubble to link this culturally rich city to each individual's vision of a better Beijing. The bubble is not regarded as a singular object, but as a means to initiate a renewed and energetic community. Under the hatchet of fast-paced development, designers must always be cognizant of Beijing's long term goals and the direction of its creativity. Perhaps designers should shift their gaze away from the attraction of new monuments and focus on the everyday lives of the city's residents.

《北京2050》描绘了三个关于北京城市未来的梦想——一个被绿色森林覆盖的天安门广场，在北京CBD上空漂浮的空中之城，和植入到四合院的胡同泡泡。经济发展所推动的大规模城市开发，正在逐步逼近北京传统的城市肌理。

面对源自城市细胞的衰退与滥用，需要从生活的层面去改变现实。并不一定要采取大尺度的重建，而是可以插入一些小尺度的元素，像磁铁一样去更新生活条件、激活邻里关系；与其他的老房子相得益彰，给各自以生命。同时这些元素应该具有繁殖的可能，在适应多种生活需求的基础上，通过改变局部的情况而达到整体社区的复苏。由此，世代生活在这里的居民可以继续快乐地生活在这里，这些元素也将成为历史的一部分，成为新陈代谢的城市细胞。

第一个"胡同泡泡32号"是一个加建的卫生间和通向屋顶平台的楼梯，它看上去仿佛是一个来自外太空的小生命体，光滑的金属曲面折射着院子里古老的建筑以及树木和天空；让历史、自然以及未来并存于一个梦幻的世界里。

胡同泡泡真正的城市理想是把北京的古城与每个人的梦想连接在一起，在大刀阔斧的城市巨变中，我们必须重新思考北京长期的目标和想象力在哪里。也许我们可以把目光的焦点从那些大型的纪念碑式建筑移开，而开始关注人们日常生活的改善和社区生活的重建。

Project name: Hutong Bubble 32
Award date: 2010
Location: Beijing, China
Building area: 130 m²
Architect: MAD Architects
Client: Beijing Zhenzheng Culture Communications Co., Ltd.
Photographer: Shu He, Fang Zhenning, Daniele Dainelli, MAD architects
Completion date: 2009
Award name: McGraw-Hill Construction 3rd Bi-Annual "Good Design Is Good Business" China Awards 2010, Best Preservation Project

项目名称：胡同泡泡32号
获奖时间：2010
项目位置：中国 北京
建筑面积：130平方米
建筑设计：MAD建筑事务所
业主：北京真正文化传媒有限公司
摄影师：舒赫 方振宁 Daniele Dainelli MAD建筑事务所
完成时间：2009
所获奖项：麦格劳－希尔公司《建筑实录》、《商业周刊》第三届"好设计创造好效益"中国奖项 2010最佳历史保护

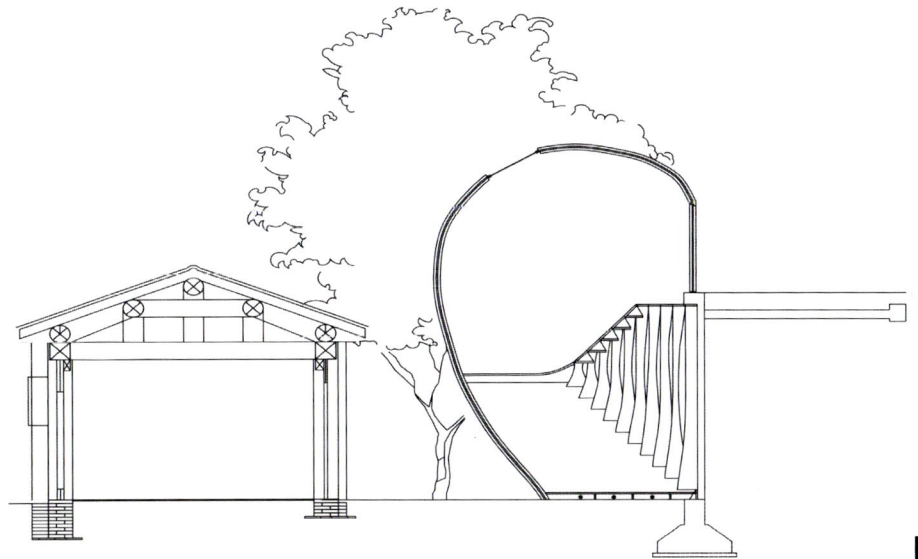

1. View from Courtyard
   庭院内部透视

2. Day View
   日景

3. Future of Hutongs
   未来胡同

4. Section
   剖面图

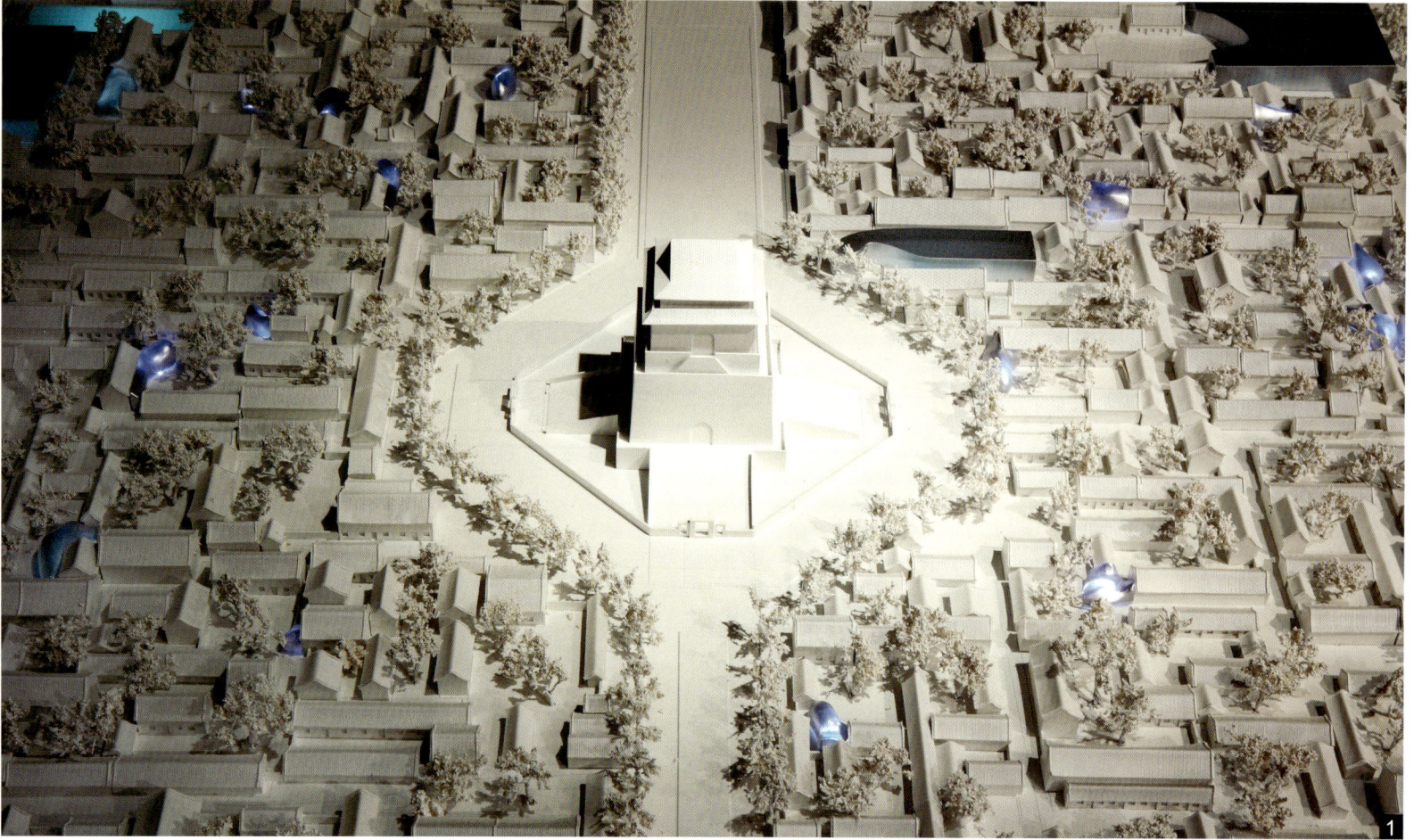

1

2

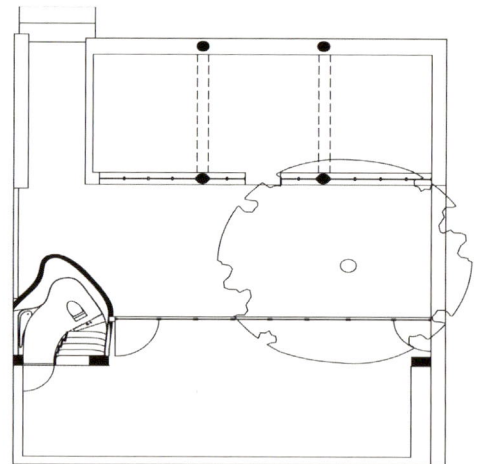

3

1. Future of Hutongs
未来胡同

2. View from Roof Terrace
屋顶平台

3. Plan
平面图

4. Interior
室内图

5. Staricase
楼梯间

6. Bathroom
卫生间

Sanghai Xiang-Dong Buddhist Art Museum

上海相东佛像艺术馆

The Museum, renovated from an old factory building, has kept remaining the roof trusses and wall textures intact, and additional spaces have been created within some areas of huge industrial space in which platforms for exhibition ascend up following the circulation route, which makes both touring and displaying together as an experience. It also introduces plants and vegetation, natural lights into the indoor space in order to create exterior effect for an interior space.

In order to express an idea of depicting the charming spirit left from the history by taking use of appropriate and reasonable materials, construction method, space and color, the renovation design has not only borrowed modeling language from ruins of Buddhist culture, but also tries to overturn this language by using elements such as red-colored concrete, formed steel as corner stopper, steel panel enclosure, reinforced bars as jungle-like fence and gravel paving, to interpret the historic taste under the contemporary context.

此馆由旧厂房改建而成，保留厂房内原屋顶桁架体系和重要墙面肌理，利用厂房大空间局部加建，展览平台沿流线层层递升，展行一体。并在室内植树造林，引入天光，营造室内空间室外化的效果。改造设计借鉴佛教文化遗址的造型语言，同时采用红色混凝土、型钢收边、钢板围合、钢筋丛栏、砾石铺地等当代材料语言加以颠覆，力图用恰当的材料、构造、空间和色彩来传达当代语境下的历史意韵。

Project name: Sanghai Xiang-Dong Buddhist Art Museum
Award date: 2010
Location: Shanghai, China
Building area: 5,078 m²
Architect: Jiakun Architects
Client: Shanghai Juyuan Asset Management Co., Ltd.
Photographer: Cai Kefei, Lve Hengzhong
Completion date: 2008
Award name: McGraw-Hill Construction 3rd Bi-Annual "Good Design Is Good Business" China Awards 2010, Best Preservation Project

项目名称：上海相东佛像艺术馆
获奖时间：2010
项目位置：中国 上海
建筑面积：5,078平方米
建筑设计：家琨建筑设计事务所
业主：上海菊园资产管理有限公司
摄影师：蔡克飞 吕恒中
完成时间：2008
所获奖项：麦格劳－希尔公司《建筑实录》、《商业周刊》第三届"好设计创造好效益"中国奖项　2010最佳历史保护

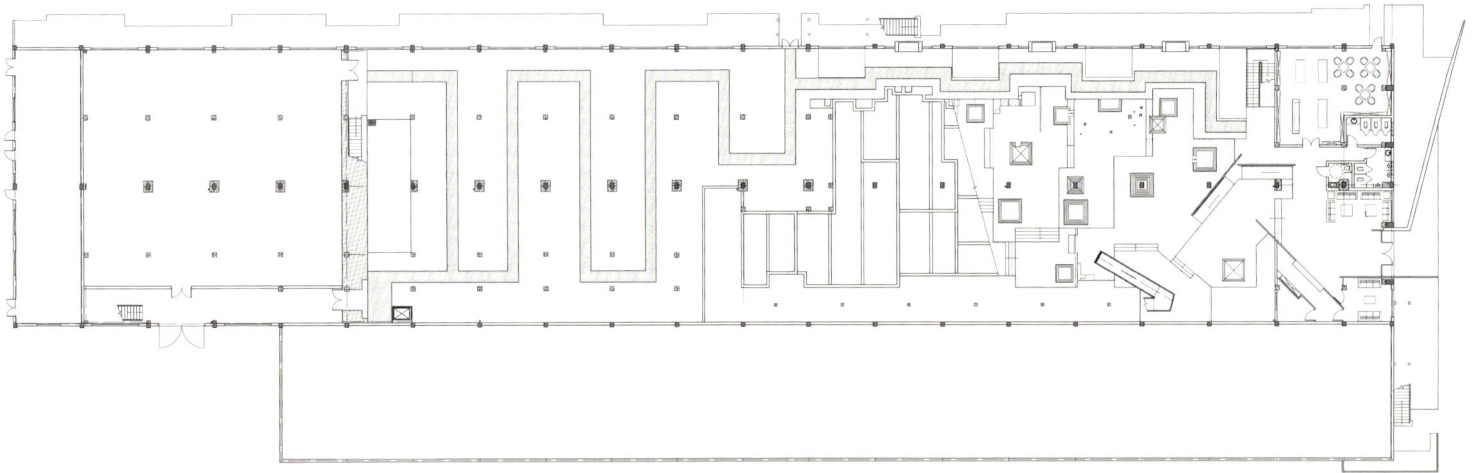

1. Exterior
   外景

2. Plan
   平面图

3. Interior
   室内

4. View before Preservation
   改造前照片

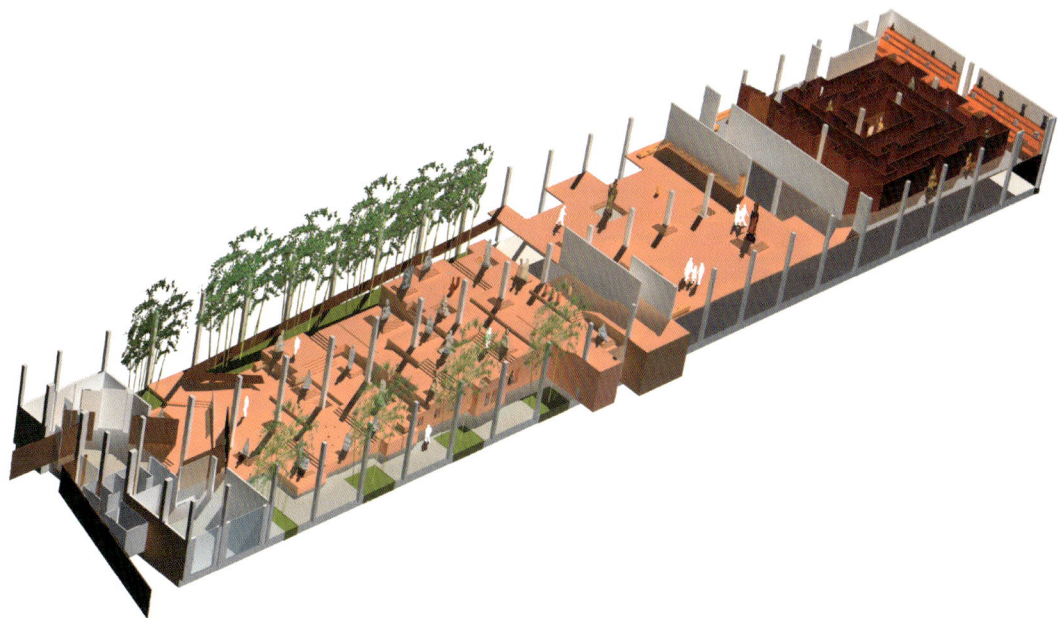

1. Interior
   室内

2. Steps
   台阶

3. Conceptual Model
   概念模型

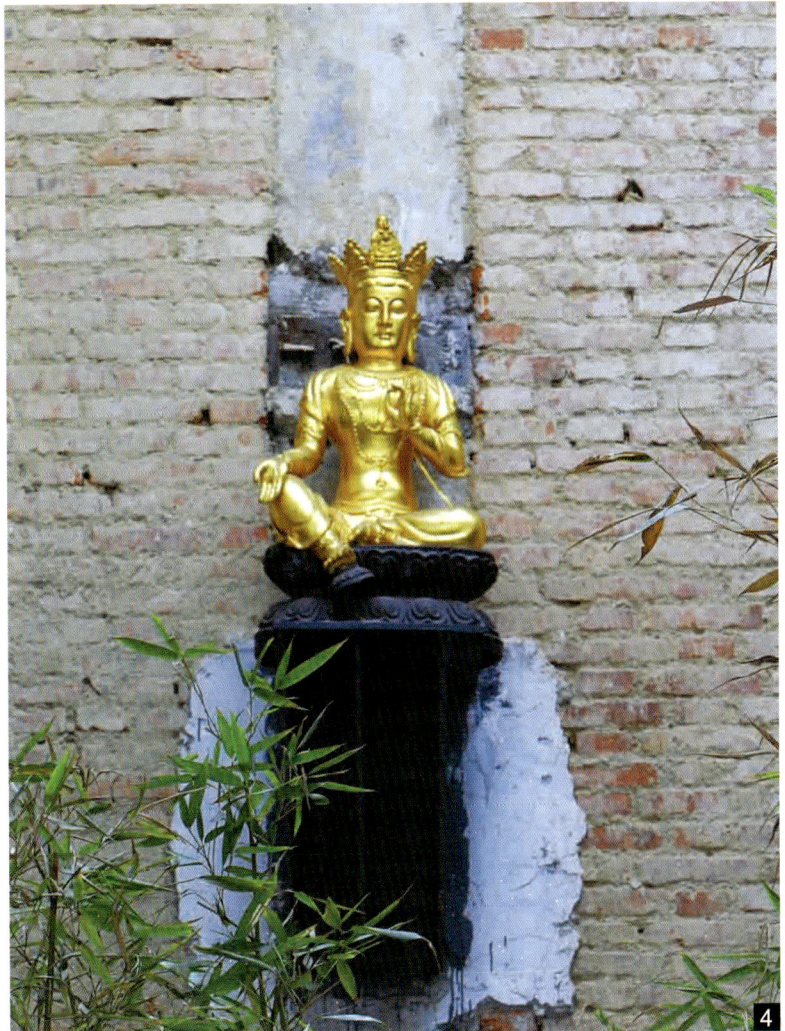

1. Display Area
   展区

2. Display Area
   展区

3. Display Area
   展区

4. Display Area
   展区

Shanghai Office of Horizon Design Co., Ltd.

群裕设计咨询（上海）有限公司

The design was initially intended to shape the office area and the space outside conference rooms into a semi-open space with natural ventilation and light. The design creates a pleasant, energy-saving micro climate that enables smooth and close interaction between people and nature.

Moreover, through many multi-purpose public spaces shaped by "architecture within architecture" like the streets and the plaza, various interpretations of the functional spaces can be given by the user. For example, spaces for discussing building material samples in the warm sunshine through the skylights, staircases to sit on and browse magazines for inspiration, a bar for exchanging thoughts on the humanistic spirit of fair trade of coffee beans, bicycling in the office areas to relieve pressure, elevated platform in the library area as a make-shift stage for rehearsing year-end party programs... Even visitors that have never visited here before can sojourn or linger freely at every corner in the public area. Through the crossing-over, various, flexible interactions, a venue atmosphere of community and campus is generated.

A large quantity of locally recycled materials is used for the project. Recyclable new materials are employed and artificially processed materials are cut down to preserve the original qualities and texture of the materials, which, besides the historical significance of extending the reuse of old buildings, carries out the eco-friendly practice of the time.

Just as Nature is the source of inspiration, most motivation of work is provided by the surrounding environment. An "office" may not be called the same in the future, but will become a kind of environment, a workplace that stimulates the worker's inner potentials, an ambience, and a spirit in enjoying the work itself.

这个项目设计之初，就是希望将办公区以及会议室以外的空间，营塑成有着自然通风、采光的"半开放"空间，借由设计手法创造出宜人舒适、节能的微气候，让人与自然有着良好而紧密的互动。

另外，借由"建筑中的建筑"围塑出许多有如街道与广场的多功能公共空间，亦可以看到使用者赋予了这些空间多重的功能诠释，例如在天窗投射的温暖阳光下讨论着建筑材料的样品、坐在阶梯上翻阅杂志寻求设计灵感、在吧台区讨论着公平交易咖啡豆的人道精神、骑自行车在办公室闲逛抒压、利用图书区高架平台当成舞台采排着尾牙的表演节目......甚至连前来参观的访客也都可以不受拘束的在公共区的每个角落游荡或逗留，借由这些跨领域、多型态、弹性的互动交流，交织出有如社区与校园的场域氛围。

在材料的选用上，大量回收再利用的当地旧材料、可以再回收的新材料以及减少过多的人为加工保留材料的原始特性与质感，除了延续旧建筑再利用的历史意义，同时也具有实践生态环保的时代意涵。

如同大自然是孕育灵感的来源，大部分的工作动力来自置身环境的供给，关于"办公室"，在未来或许没有称谓，而只是一种环境，一种可以激发内在潜能的工作场域；是一种氛围，一种乐在工作的奋起。

Project name: Shanghai Office of Horizon Design Co., Ltd.
Award date: 2008
Location: Shanghai, China
Area: 700 m²
Architect: J. J. Pan & Partners, Architects & Planners
Client: Horizon Design Co., Ltd.
Photographer: Steve Mok, Jim Chang, Lve Hengzhong
 J. J. Pan & Partners, Architects & Planners
Completion date: 2005
Award name: McGraw-Hill Construction 2nd Bi-Annual "Good Design Is Good Business" China Awards 2008, Best Historic Preservation Project

项目名称：群裕设计咨询（上海）有限公司
获奖时间：2008
项目位置：中国 上海
项目面积：700平方米
建筑设计：潘冀联合建筑师事务所
业主：群裕设计咨询（上海）有限公司
摄影师：莫尚勤　张全箴　吕恒中　潘冀联合建筑师事务所
完成时间：2005
所获奖项：麦格劳-希尔公司《建筑实录》、《商业周刊》第二届"好设计创造好效益"中国奖项 2008最佳历史保护

1. Door Way
   入口

2. Door Way
   入口

3. Outdoor Landscape
   室外景观

4. Outdoor Landscape
   室外景观

5. Outdoor Landscape
   室外景观

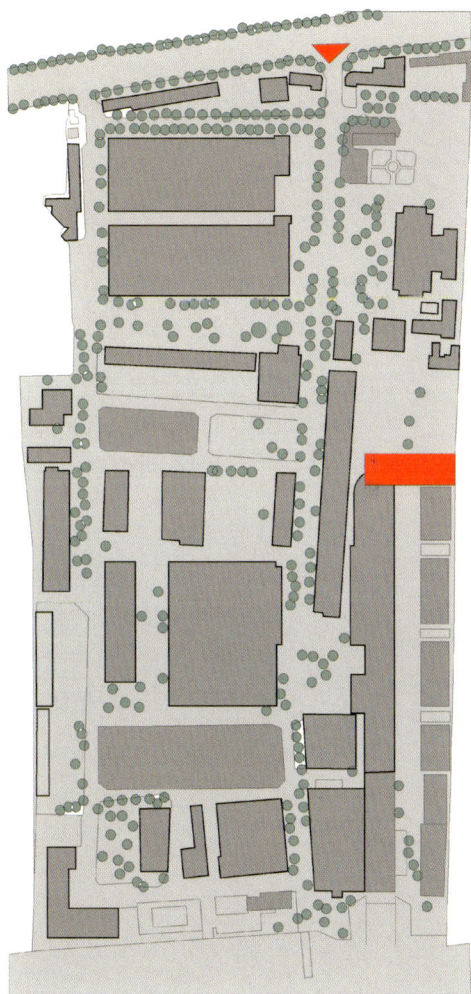

2

3

4

5

1. Lobby
   前厅

2. Site Plan
   位置平面图

3. New Office
   新办公室

4. Lounge
   休息区

5. Stairs to Lounge
   休息区台阶

1. Lobby / Exhibition Hall
   大厅/展厅
2. Conference Room
   会议室
3. Studio
   工作区
4. Lounge
   休息区
5. Sample / Model Room
   样品室
6. Administration Department
   行政部
7. Document Room
   资料室
8. Storage
   库房
9. Mechanical Room
   机房
10. Restroom / Shower Room
    卫生间/浴室
11. Security
    安全通道

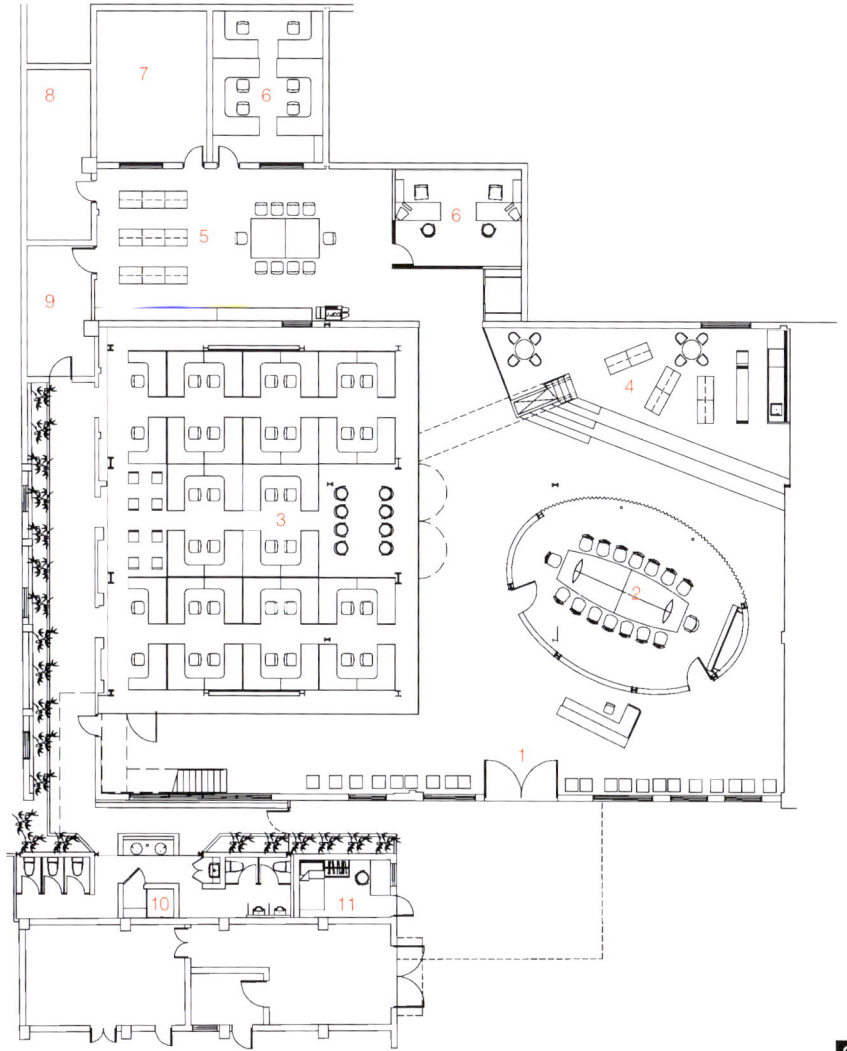

1. Conference Room Exterior
   会议室外观
2. Conference Room
   会议室
3. Plan
   平面图
4. Conference Room Interior
   会议室内部

Two to one House

"二合为一"住宅

Commissioned by a young businessman who was educated in the United States and now lives in both Hong Kong and Shanghai, Chang Bene Design took two European-style houses from the 1920s and combined them into one elegant residence. Although the architects made major changes on the inside of the buildings – creating a modern home with open, fluid spaces and contemporary style – they retained the spirit of the historic structures. Instead of trying to recreate old buildings, Chang Bene brought them alive by adapting them to modern living. The original buildings sat next to each other in the French Concession, separated by a gap 2 meters wide and 4 meters deep. To connect the buildings both horizontally and vertically, Chang Bene inserted a new stairwell/foyer tower in the gap and topped it with a right-angled skylight that brings daylight into the center of the newly combined house. The circulation tower acts as a lantern at night when it helps light up the narrow alley from which visitors enter the new house. It also offers views of a large tree outside, which helps shade the glass during the summer. The firm needed to make extensive structural alterations so that floor levels in the two buildings matched each other. Although they kept the exteriors mostly intact, the architects did enlarge some windows and repair surfaces. With the two buildings combined, the new house offers about 325 square meters of space. On the ground floor, the architects used a new dropped ceiling to create a large, unified space extending from the living room to the dining room across the width of the plan, with the kitchen and bathroom in the middle serving as a service core. Throughout the house, Chang Bene used the same gray bricks found in many other French Concession buildings, laying them with very little grout on walls, floors, and fireplaces. Upstairs, the architects opened up an unused attic above the master bedroom and converted it into a work area leading to a small roof terrace. In other places, they removed old ceilings to reveal existing wood beams.

该客户是一个曾于美国留学，现定居于香港和上海的年轻商人，希望将两个始建于20世纪20年代的欧式住宅合并成一个独立的优雅住宅建筑。设计师在保留历史建筑精髓的基础上，对原有内部空间进行了较大调整，现代、时尚的室内设计风格，为古老的建筑注入了无限生机和活力。原有两个并排建筑位于法租界，中间经由一个2米宽、4米深的空隙分隔开来。为使二者在横向和纵向上实现有效连接，设计师在这一空隙上插入了一个新的楼梯间/门厅，并在其上打造一个直角天窗，令自然光洒满整个空间。夜幕降临之时，此处仿佛一个灯笼承载着人们到达新空间。在炎热的夏日，窗外苍翠茂密的大树能够为此处进行有效遮挡。为使两个建筑间的楼层持平，设计师对建筑结构进行了调整。在尽可能保留建筑外观的同时，增设了若干窗口，并对表面进行修缮。合并后的建筑占地325平方米。在一层，设计师大胆采用了吊顶结构，打造了一个大型通透空间，穿过客厅、厨房和浴室可直达饭厅。与法租界其他建筑类似，在整个建筑中，青砖作为主要材料，分别运用到墙壁、地板以及壁炉的建设之中。楼上，设计师将主卧上方的阁楼改造成一个小型工作间，与屋顶露台相通。另外，原有天花板被彻底拆除之后，将原有的木横梁直接暴露于外。

Project name: Two to One House
Award date: 2006
Location: Shanghai, China
Building area: 320 m²
Architect: Chang Bene Design Limited
Client: C Tse.
Photographer: C Z Chai and Bao Shi Wang
Completion date: 2006
Award name: McGraw-Hill Construction 1st Bi-Annual "Good Design Is Good Business" China Awards 2006, Best Preservation Project

项目名称："二合为一"住宅
获奖时间：2006
项目位置：中国上海
建筑面积：320平方米
建筑设计：张贝理设计有限公司
业主：谢氏
摄影师：C Z Chai Bao Shi Wang
完成时间：2006
所获奖项：麦格劳－希尔公司《建筑实录》、《商业周刊》第一届"好设计创造好效益"中国奖项 2006最佳文物修复

1. Kitchen
   厨房

2. Plan
   平面图

3. Exterior
   外景

1. Garden
   花园
2. Terrace
   平台
3. Dining
   餐厅
4. Living
   起居室
5. Kitchen
   厨房
6. Foyer
   门厅
7. Breakfast
   早餐区
8. Family
   家庭区

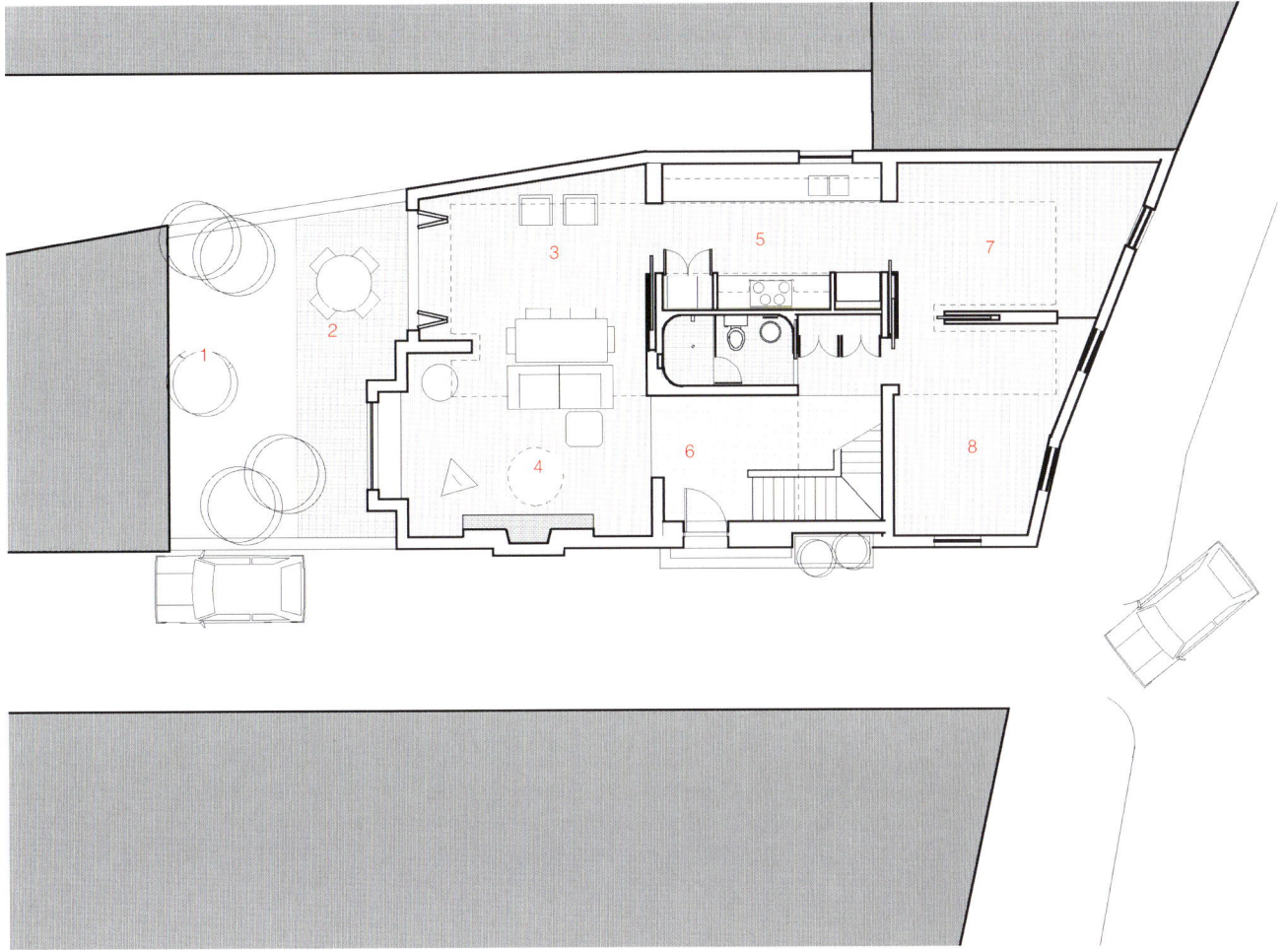

1. Master Bedroom
   主卧室

2. Master Bedroom
   主卧室

3. Foyer
   门厅

4. Bathroom
   浴室

5. Master Bathroom
   主浴室

Lijiang Ancient Town
Conservation Plan and Trust

丽江古城保护规划

With the development of tourism in the Lijiang Ancient Town since 2007, its negative impact has begun to appear: in some parts of the city, relocations have been rising up. Driven by economical benefits, the residents remodelled their houses into shops. Some run the shops themselves; some rent them to others and move out to new towns themselves. Therefore, the blocks that previously assemble residence, commerce, and tourism become tourism districts with shops. The natural trace of the history there is lost.

For the preservation and development of the Lijiang Ancient Town, considering the particular history and ethnicity there, the City Government has made many effective measures to balance the relationship between the preservation of the city heritage and the revitalization of the economy and society there.

Lijiang City Government initiated a long-term Plan of Original Residence Improvement with a donation from the Global Heritage Fund in America. The brief of the plan is to facilitate coordination between the committee and the owners to encourage the latter to reside within the old city and improve the living condition there. At present, 174 residences have been renovated and considerable social and cultural benefits have been brought out. The project won the 2007 UNESCO Asian-Pacific Region Cultural Heritage Preservation Merit Award.

1997年后随着旅游业的快速发展，其负面影响在丽江古城日渐显露：丽江古城的局部地段开始出现了不正常的迁离，居民们由于利益驱动，把住房改为店铺，或自己经营，或出租，然后迁到新城居住。这使原本集居住、商贸、游览于一体的历史街区，渐渐演变为商贸旅游区，丧失了街区的历史真实性。

对于丽江古城的保护和发展，由于丽江具有的历史和民族的特征，在充分认识到丽江古城的城市遗产保护与当地民族地区的经济、社会复兴的关系后，丽江政府已经颁布了多项措施。

丽江政府通过美国全球遗产基金会（Global Heritage Fund）的捐款作为启动资金，建立了一个比较长期稳定的"原住民住房改善计划"，这个计划需要在委员会和业主之间通过协议的方式，鼓励当地的居民生活居住在古城区内，并且改善他们的居住环境。目前，已经完成了174幢民居的修缮补助计划，并且取得了非常好的社会、文化效益，获得了2007年"联合国教科文组织亚太地区文化遗产保护优秀奖"。

Project name: Lijiang Ancient Town Conservation Plan and Trust
Award date: 2006
Location: Yunnan, China
Planned area: 3.8 km²
Architect: Shanghai Tongji Urban Planning & Design Institute
Client: Lijiang Ancient Town/UNESCO World Heritage Center
Completion date: 2006
Award name: McGraw-Hill Construction 1st Bi-Annual "Good Design Is Good Business" China Awards 2006, Best Preservation Project

项目名称：丽江古城保护规划
获奖时间：2006
项目位置：中国 云南
规划面积：3.8平方公里
规划设计：上海同济城市规划设计研究院
业主：全球遗产基金会，丽江古城管理委员会，联合国科教文组织世界遗产中心
完成时间：2006
所获奖项：麦格劳-希尔公司《建筑实录》、《商业周刊》第一届"好设计创造好效益"中国奖项 2006最佳文物修复

1. Scenery
   秀美风光

2. Topography
   地形图

3. Bird View
   鸟瞰

4. Before Preservation
   修缮前

5. After Preservation
   修缮后

1

2

1. Structure
   结构分析图

2. After Preservation
   修缮后

3. After Preservation
   修缮后

4. After Preservation
   修缮后

5. Plans of the Traditional
   Layouts
   传统民居分布图

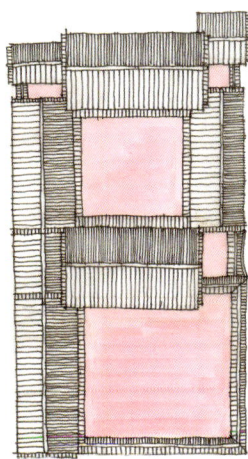

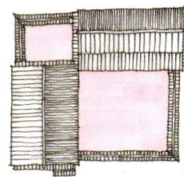

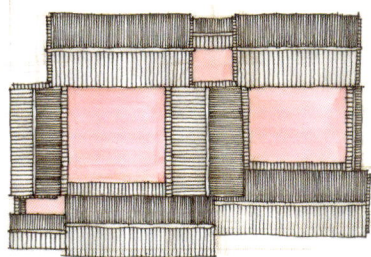

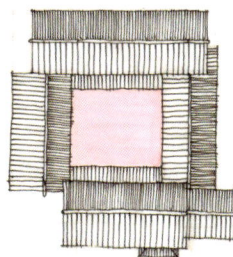

Jianfu Palace Garden

建福宫花园

The Garden of the Palace of Established Happiness (Jianfu Palace Garden) was built by the Emperor Qianlong in 1740 and is composed of a series of pavilions set in garden courts in the northwest corner of the Forbidden City. Destroyed by a mysterious fire in 1923, the site was left vacant for over 75 years. In a 5-year collaboration between the Hong Kong-based China Heritage Fund and the Palace Museum, the entire complex was painstakingly reconstructed by master craftsmen, carpenters, masons, tile workers and painters who worked together using traditional tools, techniques and processes.

China Heritage Fund's aim is to revive traditional building crafts as well as the training of artisans, as a means of conserving China's rich cultural past. The US-based Tsao & Mckown Architects together with Pei Partnership Architects have transformed the reconstructed interiors of the complex into a series of exhibition, reception and meeting spaces for special visitors to the Palace Museum. State-of-the-art mechanical and electrical systems have been inserted into the traditional pavilions in a discreet and respectful manner to ensure that their provision did not jeopardize the delicate wooden structures. The interiors of the pavilions have been left exposed as much as possible so as to illustrate the sophistication and intricacy of traditional Chinese imperial architectural techniques.

At the same time, there is a richness and sumptuousness in textures and furnishings, employing both antiques from the Palace Museum Archives and reinterpreted furniture and accessories, to satisfy the requirements of the pavilions' new uses. The project was completed in the autumn of 2005, in time for the 80th anniversary of the Palace Museum.

建福宫花园，位于紫禁城西北隅，始建于清乾隆5年（1740年），面积4,000多平方米。作为皇帝休闲放松的场所，建福宫花园院落层次分明，是紫禁城内空间变化最丰富的院落，也是融皇家园林与江南园林于一身的佳作。

1923年6月26日，一场神秘大火将整个花园连同无数珍宝一夜化为灰烬。自此，建福宫花园废墟沉睡在瓦砾之下长达75年之久。20世纪初，借着香港中国文物保护基金会的捐助与鼎力支持，这座气派非凡的皇家园林经故宫博物院的木、石、瓦与油饰彩画等能工巧匠五年多的努力，采用传统工具和手艺进行复建，重现出昔日的风采。重建目的亦包括复兴中国传统楼宇的修建和工匠们的传统技艺，从而将之作为保存和发扬中国富有的历史文化遗产的一种途径。

故宫博物院将复建的建福宫花园其功能定位为接待各国贵宾、举办讲座、招待会和兴办特别展览等文化活动场所。香港中国文物保护基金会聘请了Tsao & McKown室内设计事务所与贝氏建筑师事务所承担室内装修设计，设计师们既尊重和展示这组皇家建筑自身的建筑美和结构美，又考虑和当今的实用功能相吻合，将空调等现代技术应用于其中而不破坏原有建筑结构，实现了皇家园林与现代设施的实用完美结合。这次修缮于2005年秋天完成，当时正值故宫博物馆的80周年纪念。

Project name: Jianfu Palace Garden
Award date: 2006
Location: Forbidden City, Beijing, China
Building area: ~ 2,800 m²
Architect: The Palace Museum Historical Architecture Conservation Center
Principal Design Consultants: Tsao & McKown Architects, with Pei Partnership Architects
Client: China Heritage Fund
Photographer: Cheng Shouqi
Completion date: 2005
Award name: McGraw-Hill Construction 1st Bi-Annual "Good Design Is Good Business" China Awards 2006, Best Historic Preservation Project

项目名称：建福宫花园
获奖时间：2006
项目位置：北京紫禁城
建筑面积：约2800平米
古建筑师：故宫博物院古建修缮中心
建筑师：Tsao & McKown Architects,贝氏建筑师事务所
业主：故宫博物院，中国文物保护基金会
摄影师：程受琦
完成时间：2005
所获奖项：麦格劳-希尔公司《建筑实录》、《商业周刊》第一届"好设计创造好效益"中国奖项 2006最佳文物修复

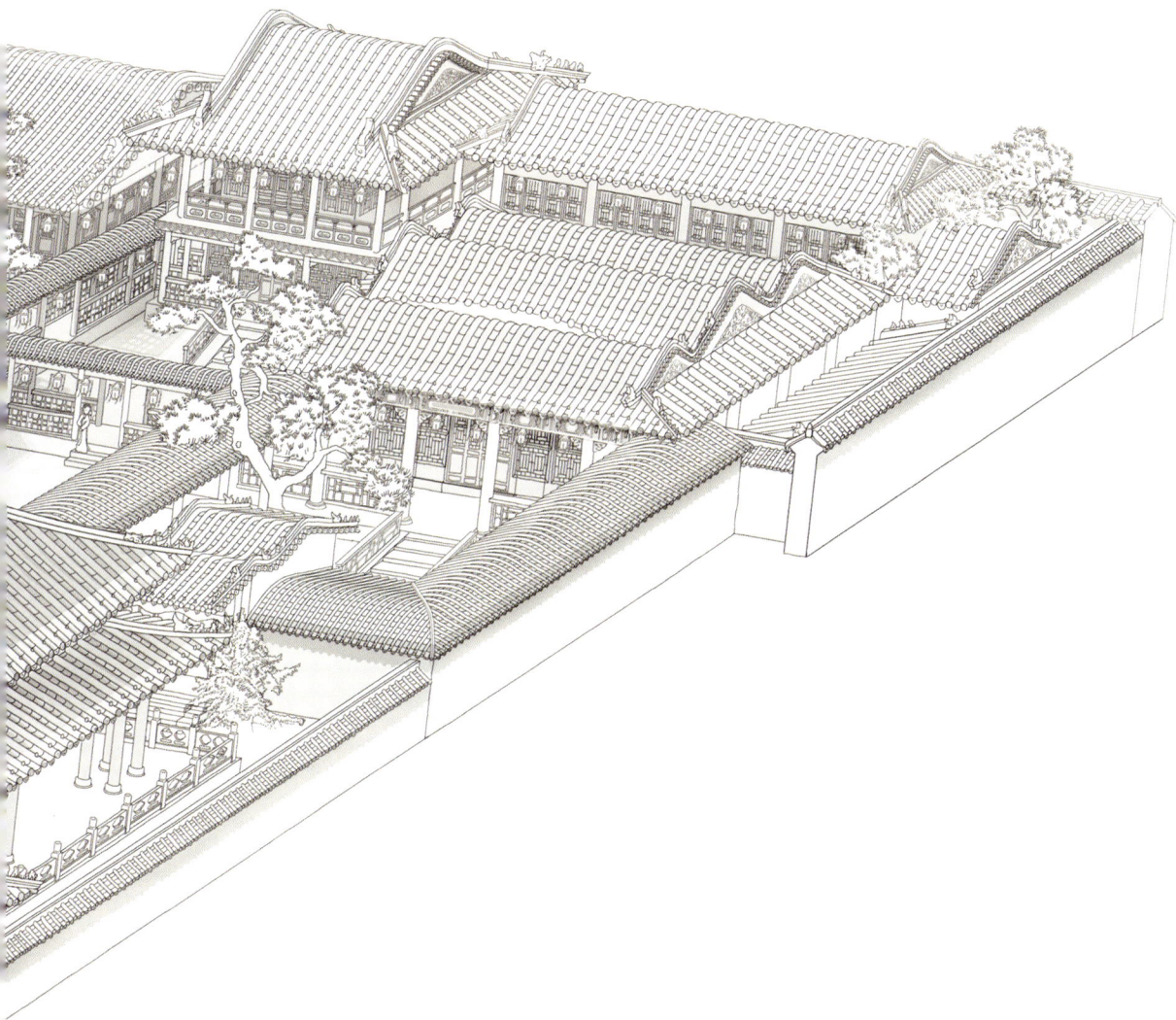

1. Rendering of the Jianfu Palace & the Jianfu Palace Garden. Drawing by Zhao Guangchao
   建福宫及建福宫花园示意图。赵广超绘制

2. Jianfu Palace Garden Site before Reconstruction
   建福宫花园遗址

1. Entrance to the Jing Sheng Zhai Pavilion
   敬胜斋入口

2. Location of the Jianfu Palace Garden
   in the Forbidden City
   建福宫花园方位图

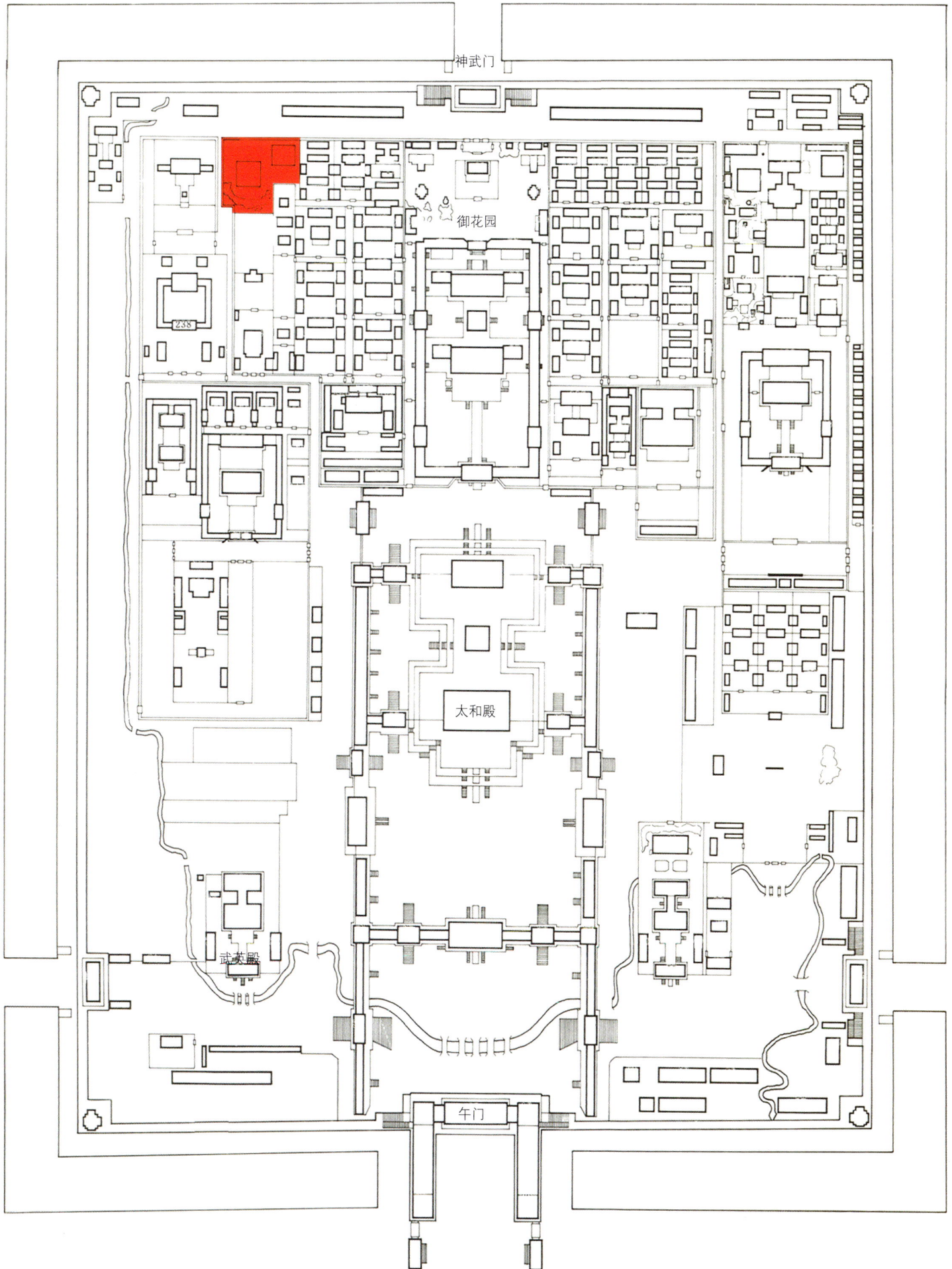

1

2

1. Interior View of the Jing Sheng Zhai
   Pavilion
   敬胜斋内景

2. Interior View of the Jing Sheng Zhai
   Pavilion
   敬胜斋内景

3. Interior View of the Ji Yun Lou Pavilion
   吉云楼内景

4. Ground Floor of the Yan Chun Ge
   Pavilion
   延春阁首层内景

5. Jianfu Palace Garden Site Plan
   建福宫花园总平面图

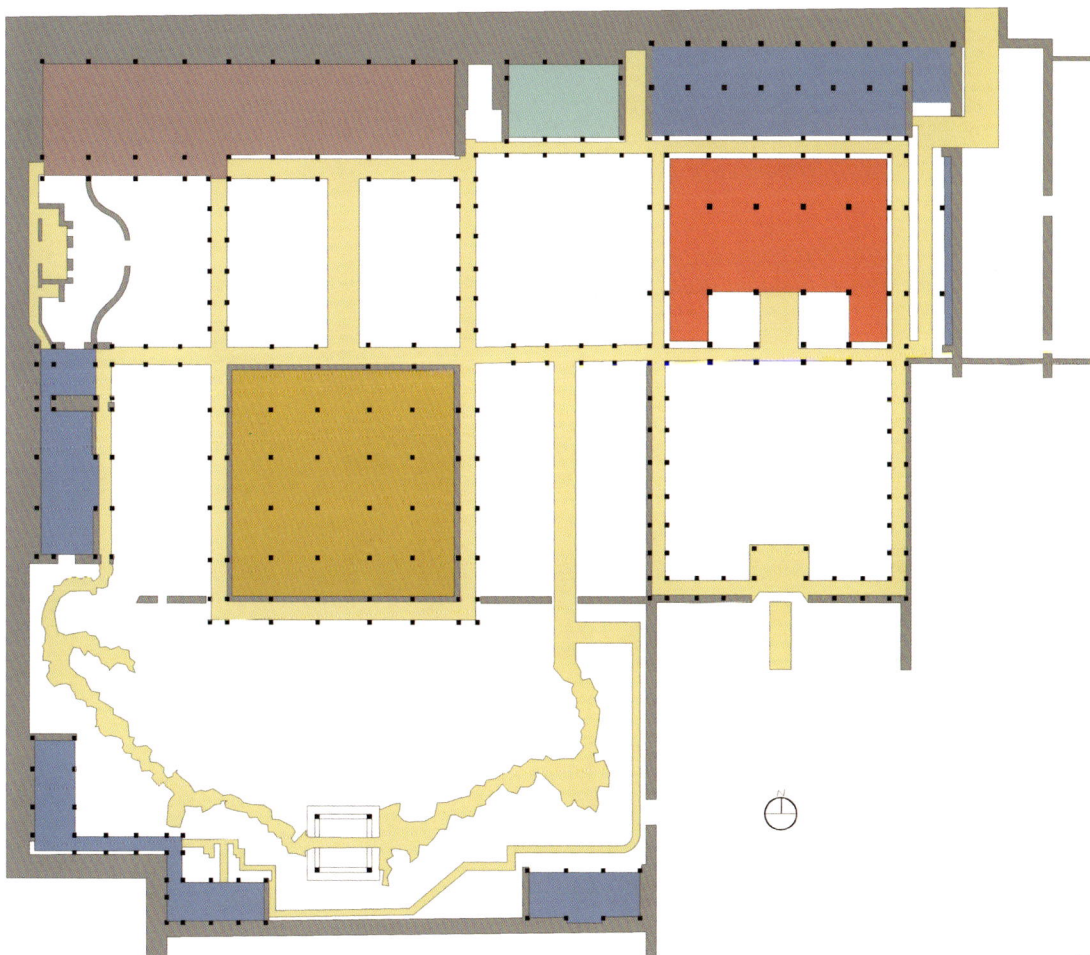

1. Yan Chun Ge
   延春阁
2. Jing Yi Xuan
   静怡轩
3. Jing Sheng Zhai
   敬胜斋
4. Ji Yun Lou
   吉云楼
5. Administration, Storage & Guardhouse
   行政楼、库房及安保室

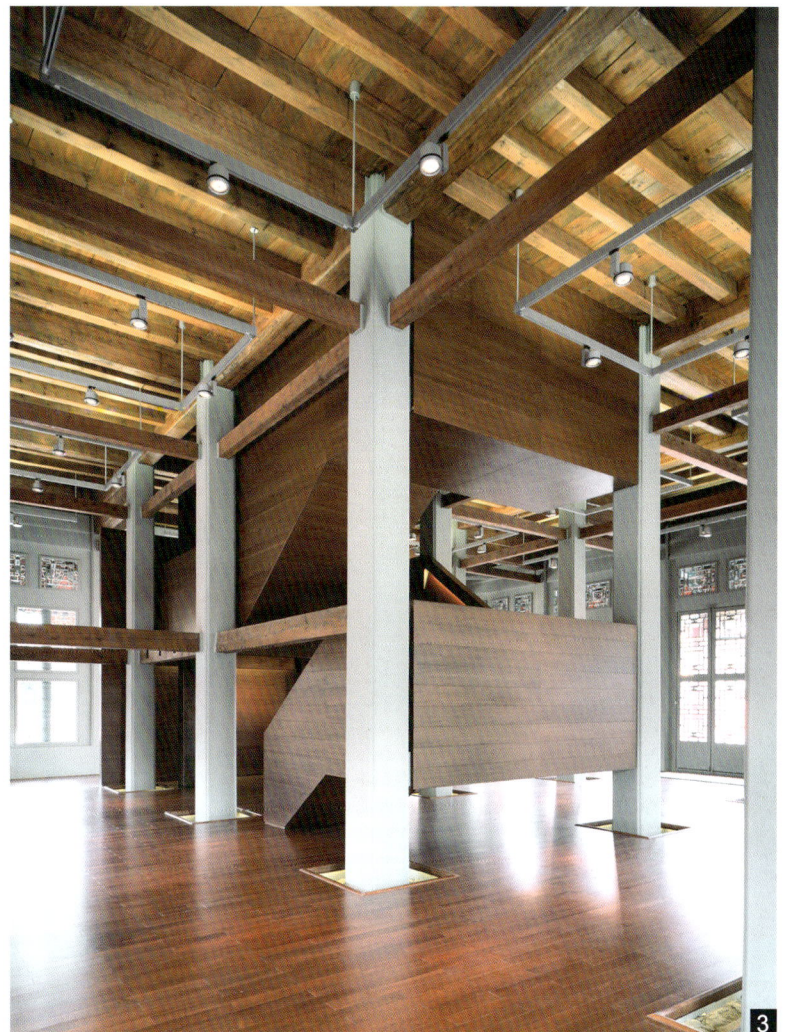

1. View of the Third Floor of the Yan Chun Ge
   Pavilion, with the Central Staircase and the
   Exposed Wood Roof Structure
   延春阁顶层内景

2. A New Staircase Inserted into the Wood
   Structure of the Yan Chun Ge Pavilion
   延春阁新楼梯

3. View from the Staircase in the Yan Chun Ge
   Pavilion
   延春阁首层内景

Longchi Town Conceptual Master Plan

龙池镇概念规划

It seemed very difficult to rebuild Longchi on the ruins left by the 2008 earthquake at the beginning. It was in urgent need to provide dwellings for 2,000 farmers as soon as possible and set up a new multi functional community. Taking the topography, water and other environmental concerns into consideration, only about 55 ha of the land is available.

Zooming out from the small valley we get a bigger picture of the region, where lots of tourism resources are there to be hold together. Being connected to resources from larger area brings opportunity for Longchi Town to be developed as a regional tourism service center. Instead of separating tourists and local residents like many other tourist destinations in China, the designer tries to mingle the two, allowing them to share some of the public spaces / amenities, and let them benefit from each other. On the southern bank the relocation community is connected with the waterfront pedestrian street open to tourists. A series of local style mountain courtyard provides 15,000 sqm retail space for local residents to run family business. At the end of the pedestrian street a hotel / conference center with earthquake memorial hall is located together with residential community center sharing the same public walkway / plaza.

A sensible preservation of environmentally sensitive land gives opportunity to mixed use. By providing good infrastructure and smart public space it is hoped to create a tourist town that is authentically living. The continuous interaction between the locals and visitors will make this project truly sustainable socially and economically.

龙池，遭遇了2008年四川大地震之后成为一片废墟。重建的起步似乎很难。在这样一个小山谷中，不但需要尽快解决2000户农民的居住问题，同时需要构建一个集居住、娱乐等于一体的多功能新型社区。排除地形、水源及其环境的干扰，实际上大约只有55公项土地可用来构建。

从龙池的上空拍下来的照片来看，龙池拥有很多丰富的旅游资源。庞大的旅游资源使龙池镇逐步发展成为一个特色旅游服务中心。在该项目中，和以往的旅游景点不同，设计师并未将游客与当地居民分离开来，而是使二者合二为一，即使他们共同分享一些公共场所和设施，使双方均受益。南岸重建的社区内与海滨步行街相通，对游客开放。15,000平方米的销售区供当地居民经营各自的"农家乐"。在步行街的尽头，设有地震纪念馆酒店/会议中心以及住宅社区中心，二者分享一个通道和广场。

该项目的成功之处在于对原有环境的保留以及充分利用。通过创建良好的基础设施和智能化公共空间，打造一个精巧，更具文化内涵的旅游小镇。当地居民和游客之间的互动将大大促进该项目社会化和经济化的可持续发展。

Project name: Longchi Town Conceptual Master Plan
Award date: 2010
Location: Chengdu, China
Planned area: 110 hectares
Architect: AECOM
Client: Chengdu Culture & Tourism Development Group LLC
Completion date: 2009
Award name: McGraw-Hill Construction 3rd Bi-Annual "Good Design Is Good Business" China Awards 2010, Best Planning Project

项目名称：龙池镇概念规划
获奖时间：2010
项目位置：中国 成都
规划面积：110公项
规划设计：AECOM
业主：成都文化旅游发展集团有限责任公司
完成时间：2009
所获奖项：麦格劳–希尔公司《建筑实录》、《商业周刊》第三届"好设计创造好效益"中国奖项　2010最佳规划设计

1

2

3

1. Illustrative Master Plan
   总平面图

2. Culture & Conference Center
   3D Model
   文化会议中心3D模型

3. Overall Bird Eye View
   总体鸟瞰图

1. Elevation of Waterfront Commercial Street
滨水商业街立面

2. Rendering of Waterfront Commercial Street
滨水商业街效果图

3. Rendering of Water Plaza
亲水广场效果图

Meixi Lake Master Plan
梅溪湖总体规划

The Meixi Lake master plan establishes a paradigm of man living in balance with nature. The densely concentrated urban plan, packed with a full variety of functions and building types, is integrated with mountains, lakes, parks and canals, resulting in an environment which promotes both health and prosperity. As a new center within the larger metropolitan area of Changsha, Meixi Lake offers a new model for the future of the Chinese city. Advanced environmental engineering, pedestrian planning, cluster zoning, and garden integration are all part of a holistic design strategy in this healthy city.

The first element of the Meixi plan is water. The mixed-use central business district is wrapped around the circular heart of this body of water. In this district, high-rise building districts are connected by a pedestrian tram street, reducing the need for car use in the city center. A series of canals radiate from the water's edge, allowing for boat transport from the city center to any one of eight neighborhood clusters. Each cluster houses about 10,000 people, and includes a village center featuring a school, shopping area, and other spaces for public use. These neighborhoods are separated from one another by green buffers which accommodate exercise fields and natural landscape zones.

The radial geometry of the city plan allows for a highly efficient transport system, reducing potential pollution and energy use. Other environmental strategies include collective gray and black water systems, distributed energy plants, and urban agriculture. A river flood plane is turned into a linear park which includes recreational areas, micro farms, and residential rows.

Overall, the design of Meixi allows the vitality of a dense metropolis to be combined with the beneficial qualities of a natural setting. This forward-looking community will benefit from and promote the development of new technologies. Both a major convention center and an Education/R+D sector will encourage Chinese and foreign businesses to consider Meixi Lake an ideal place to demonstrate new ideas about the way we live.

梅溪湖总体规划旨在探索树立人与自然和谐相处的典范。高度集中的城镇规划、齐全的配套设施与多样的建筑类型，与山脉、湖泊、公园以及运河结合在一起，形成了促进健康和繁荣的良好环境；作为长沙大都市范围内的一个新中心，梅溪湖规划为未来的中国城市建设树立了一个新的典范；领先的环境设计、人行道规划、功能分区以及绿化的融合，所有这一切都构成了这座健康城市的总体设计战略。

梅溪湖的第一要素是水。多元化的中央商务区在梅溪湖畔依水而建。在这里，有轨电车和步行街将高楼区连接在一起，在市中心可以大大减少对机动车的使用频率。自湖边辐射出的是一系列的运河，这些运河使得船舶可以从市中心直达八个相邻社区中的任意一个；每个社区可以容纳大约10000居民，并包含由社区学校、购物区和其他的公共设施组成的社区中心。邻近的社区之间通过绿色缓冲地带隔开，缓冲地带内设有健身场地和天然景观带。放射状几何布局的城市规划能够实现城市交通系统高速便捷、降低潜在的污染和能源消耗；其他的环境战略包括集中的灰黑水处理系统、分散式的能源设备和都市绿野；将河道防洪堤变为包括娱乐休闲区域、小型农田和居住带的线性公园。

总而言之，梅溪湖的设计使得密集而充满活力的大都市能够与高品质的自然景观融合在一起；这种前瞻性的社区将会从新科技中受益，同时又能促进新科技的发展；大型会展中心与教育研发部门两者都将会激励中外商家将该区域作为展示新型生活理念的理想之地。

Project name: Meixi Lake Master Plan
Award date: 2010
Location: Changsha, China
Planned area: 645 hectares
Architect: KPF Associates PC
Client: City of Changsha
Award name: McGraw-Hill Construction 3rd Bi-Annual "Good Design Is Good Business" China Awards 2010, Best Planning Project

项目名称：梅溪湖总体规划
获奖时间：2010
项目位置：中国 长沙
规划面积：645公顷
规划设计：KPF建筑师事务所
业主：长沙市人民政府
所获奖项：麦格劳希尔公司《建筑实录》、《商业周刊》第三届"好设计创造好效益"中国奖项 2010最佳规划设计

1

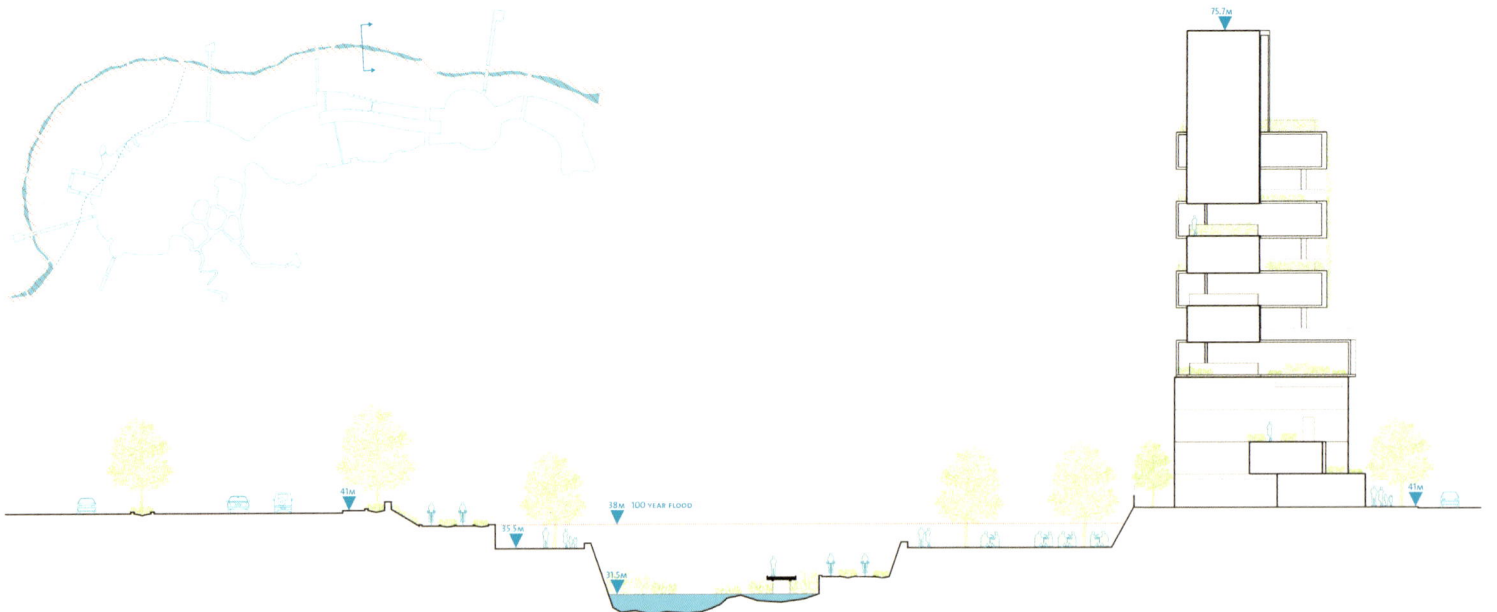

75.7м

2

1. Garden City Festival Island
   花园城市：节庆岛屿

2. Drawing
   图纸

3. Convention Center
   会展中心

4. CBD City Island
   中央商务区：城市岛屿

3

4

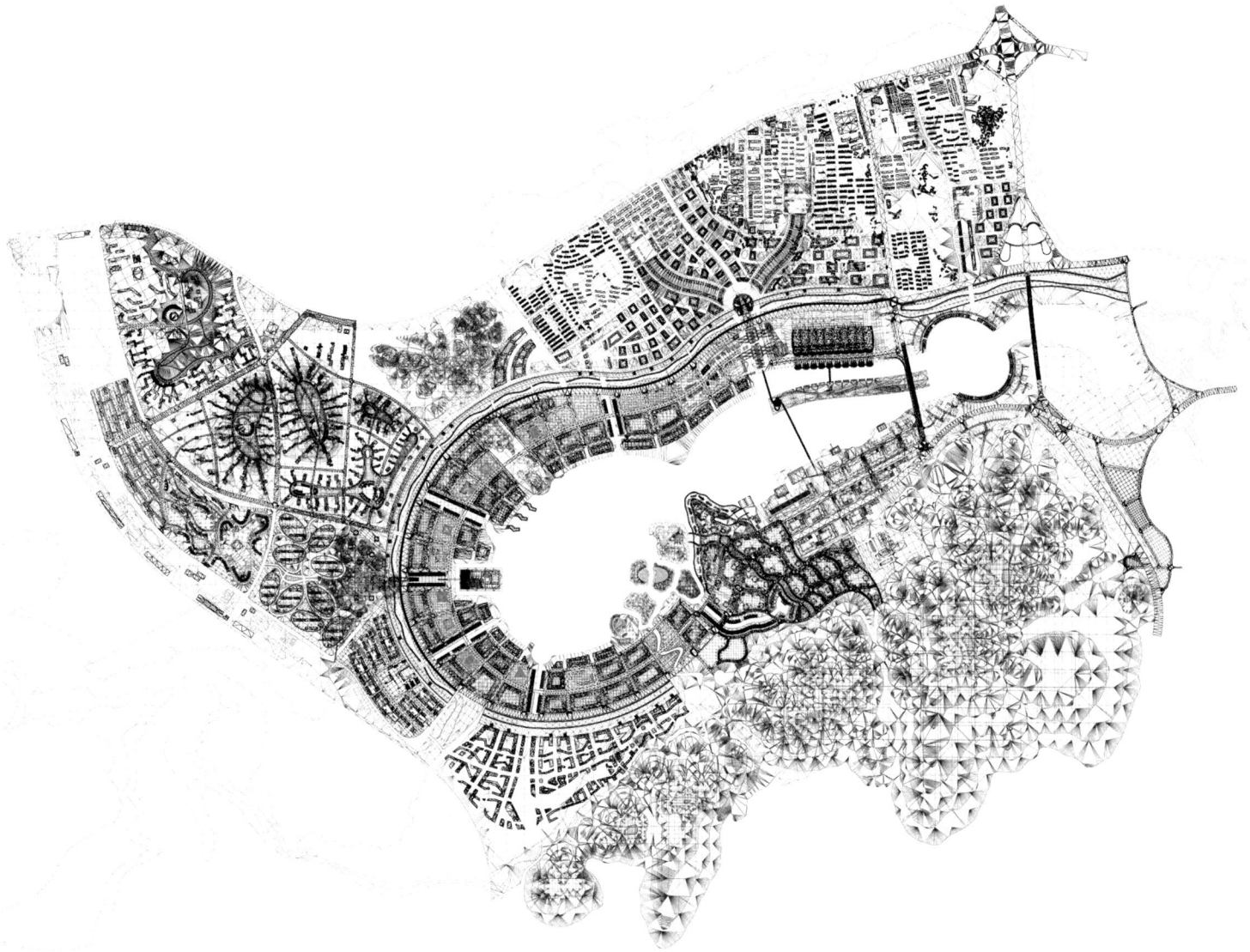

2

1. CBD Urban Planning
   中央商务区城市规划

2. Plan
   平面图

Beijing Financial Street

北京金融街中心区

Located five blocks west of the Forbidden City, Beijing Finance Street (BFS) is a 10-million-square-foot (860,000-sm) mixed-use district that seeks to establish a new form of modern urbanism in China. Throughout much of the country's contemporary urban development, the pervasion of superblocks and wide setbacks has tended to result in massive buildings that disengage from the street and each other. Rather than working together to create a place, these buildings function as stand-alone, mixed-use islands that lack a sufficient amount of publicly oriented uses to establish a larger presence or community hub.

A Model of Sustainable Redevelopment for China

As a unique center of commerce with a diverse mix of tenants and uses, BFS serves as China's new "Wall Street". With its neighborhood-serving and destination retail components, conference center, hotels and commercial space, residential neighborhoods, and connected gardens and open spaces, BFS provides an enriching center for the larger Finance Street district that is already positively influencing subsequent development.

A Place for People, Served by Transit

The centerpiece of BFS is an eight-acre (three-hectare) civic park that is activated by cafes, restaurants, and retail along its perimeter. Unlike other gardens in Beijing that are walled and have controlled points of entry, the park at BFS is a highly permeable open space that is enlivened by the varied mix of people that populate it from early morning until late at night. The park's design preserves all of the historic trees on the site and adds more than 500 new gingko trees that give BFS its seasonal character. To minimize potable water consumption, native plants are used throughout the park and gray water irrigates the landscape.

Environmentally Responsive Design

The 22 buildings of BFS are organized around seven south-facing courtyards, with contoured massing to allow maximum sunlight exposure for dwelling units and open spaces. The residential buildings on the south sides of the blocks bend in plan to follow the sinuous perimeter of the park and remain low to preserve an intimate connection with the landscape. A series of north-south and east-west public walkways make the blocks porous by penetrating through the building massing and linking the internal courtyards both to the central park and to the city beyond.

北京金融街（BFS）是一个新型都市多功能区，共分为五个部分，位于紫禁城的西侧，面积达860,000平方米。随着都市现代化进程的加快，摩天大楼的不断兴起，大型建筑群落正悄然形成。在各独立的建筑间巧妙建立一种联系，打造大型多功能社区成为当前的热点。

中国可持续发展规划的范例

作为一个多功能商业中心，BFS在中国扮演了"华尔街"的角色。其中设有零售店、会议中心、酒店与商业空间、住宅区、花园以及公共空间，BFS的创建将大大促进城市更长远的发展。

优良的人居环境，交通便利

BFS的核心是一个占地3公顷的市民公园，其周边设有咖啡馆、餐厅和零售店。通常，公园都是用围墙进行圈定之后对出入口加以限制，与此相反，该市民公园是一个更为开放的空间，对市民进行全天开放。在设计过程中，百年古树被保留在一侧，而另一侧则增植了500多棵银杏树，令整个公园四季风采各异。为了尽量减少水源的消耗，公园中还种植了大量原生植物，景观采用灰水进行灌溉。

响应环保设计

BFS中的22个建筑以7个南向的庭院为中心呈波浪形布局，便于住宅区和公共区的采光。社区的南端是住宅区，呈弧形设计，与蜿蜒的公园外围自然融为一体，体现了人文与自然结合的设计理念。若干条南北和东西向公共通道将建筑与庭院乃至中央公园以及整个城市完美衔接起来。

Project name: Beijing Financial Street
Awarded date: 2008
Location: Beijing, China
Site area: 1,030,000 m²
Architect: Skidmore, Owings & Merrill LLP
Client: Beijing Finance Street Holding Co
Photographer: Courtesy of Skidmore, Owings& Merrill LLP, Tim Griffith
Completion date: 2007
Award name: McGraw-Hill Construction 2nd Bi-Annual "Good Design Is Good Business" China Awards 2008, Best Planning Project

项目名称：北京金融街
获奖时间：2008
项目位置：中国 北京
占地面积：1,030,000平方米
规划设计：SOM建筑设计公司
业主：北京金融街控股股份有限公司
摄影师：SOM建筑设计公司　Tim Griffith
完成时间：2007
奖项名称：麦格劳-希尔公司《建筑实录》、《商业周刊》第二届"好设计创造好效益"中国奖项　2008最佳规划设计

1. Building Exterior
   建筑外景

2. Plan
   平面图

3. View from the Street
   街景

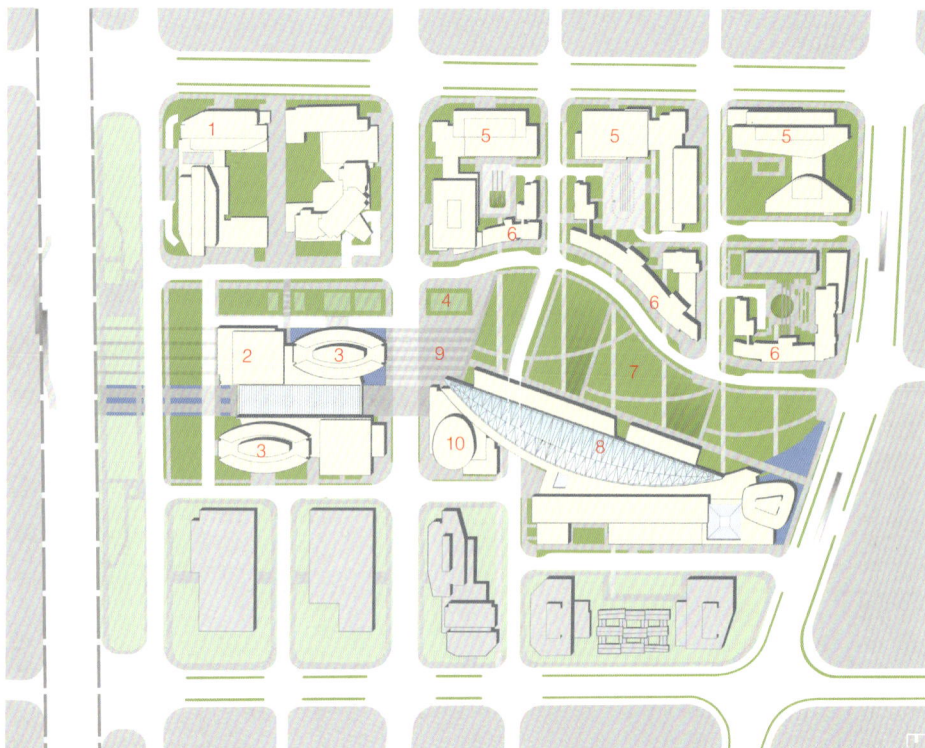

1. China Insurance Regulatory Commission
   中国保险监督管理委员会

2. Conference Hall
   会议中心

3. Landmark Tower
   地标建筑

4. Finance Plaza
   金融广场

5. Office Tower
   办公楼

6. Low Scale Residential
   低密度居住区

7. Central Park
   中央公园

8. Retail Atrium
   零售区

9. Fountain
   喷泉

10. Theater
    剧场

3

1. Night View
   夜景

2. Entrance of China Insurance Regulatory Commission
   中国保监会入口

3. Fountain of the Street
   街心喷泉

4. Walkpath
   步行道

5. Interior of the building
   建筑室内

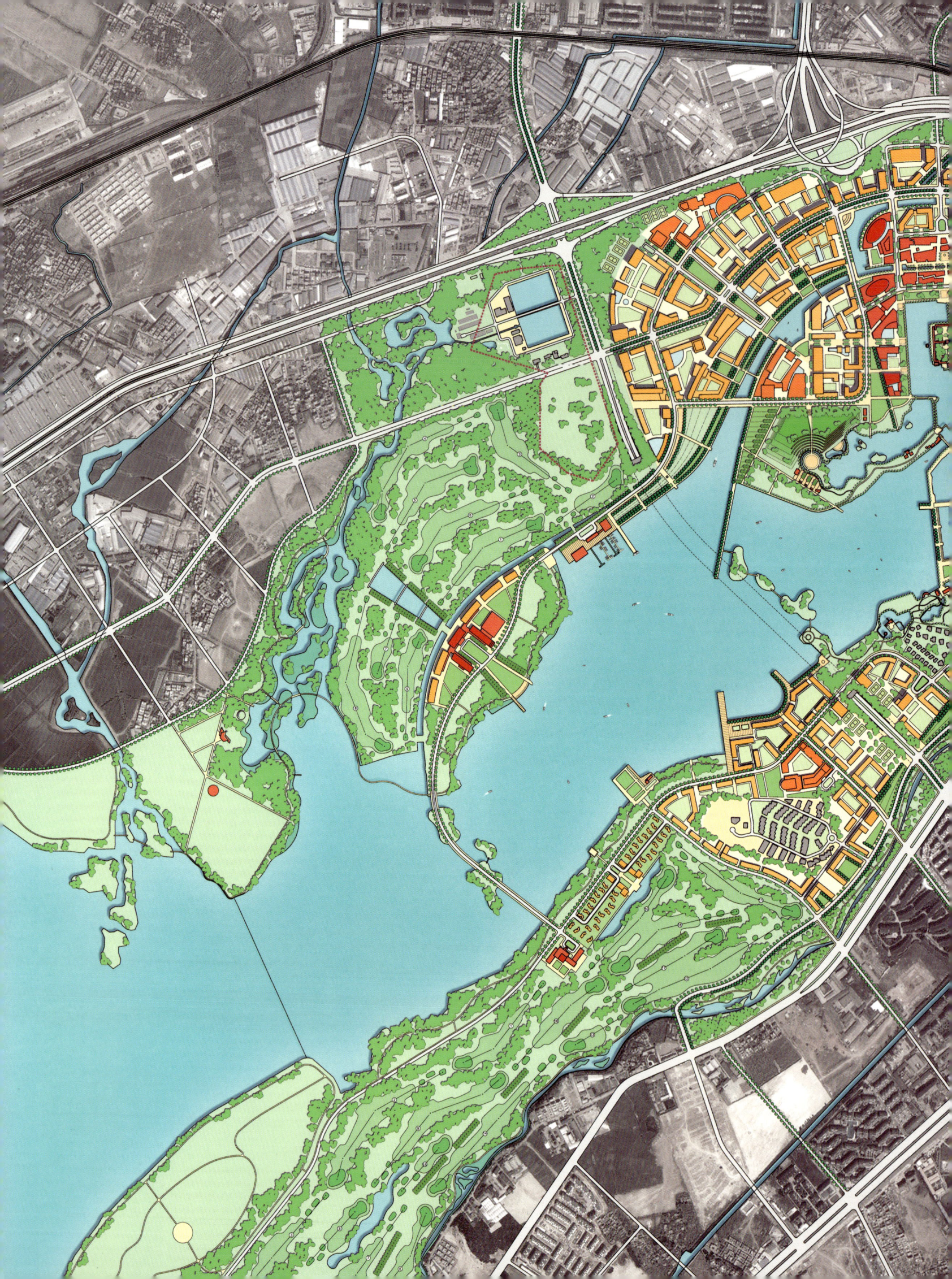

Caohai North Shore Conceptual Master Plan

草海北岸概念性总体规划

Known as China's "Eternal Spring City", Kunming has a year-round temperate climate and picturesque scenery. It is the gateway to China for members of the Association of Southeast Asian Nations (ASEAN) such as Thailand and Vietnam. This dynamic growth is leading Kunming to quickly evolve from a quiet, provincial capital into an international destination for business and tourism.

Working with Shui On Land, Sasaki helped to identify approximately 520 hectares of land, which would reconnect the core of the city to the shores of Dianchi Lake. The urban-design strategy seeks to transform this previously neglected waterfront into the new "living room" of Kunming, composed of cultural and entertainment venues, pedestrian-oriented streets with ground level retail and restaurants, a collection of offices geared toward creative industry, multiple schools, and a variety of unique residential neighborhoods with dramatic views of the lake.

Caohai North Shore is designed as a sustainable new community which can become a model for ecologically restorative development in the region. One of the key elements of the plan is to restore a portion of the lake (currently one of the most polluted bodies of water in China) via strategies which identify locations for new wastewater treatment plants, dredge existing sediment to reduce current phosphorus and nitrogen loads, construct additional wetlands to filter stormwater from the community, and re-introduce native vegetation and wildlife.

In addition to re-establishing the ecological diversity and overall health of the lake, the plan incorporates other environmental principles. The compact community plan provides efficient connections to public transit, creates a pedestrian-oriented network of streets, and protects and enhances open space. One of the most significant contributions to the public realm is the renovation of Daguan Park. Respecting the park's 800-year history, its traditional core will be refurbished, and new open space will be used for a variety of recreational and educational activities. Adding to the existing unique qualities of the region, the architecture of the community is designed as a contemporary interpretation of the traditional local villages. The building orientation takes advantage of lake breezes for natural ventilation, and generous parks and plazas encourage use of outdoor space.

作为"中国春城"，昆明一年四季都有着温暖的气候和如画的风景。她是东南亚国家例如泰国和越南等进入中国的门户。这种动态的成长使昆明从一个安静的省会城市迅速发展为一个国际化的商业和旅游中心。

与瑞安公司一起，Sasaki 协助规划了大约520公顷的土地，将昆明的核心地区与滇池连接到一起。城市规划的战略要求改变之前被忽视的水岸，从而形成新的"昆明前沿"，发展由文化与娱乐场所，遍布零售店和餐厅的步行商业街以及办公区域组成的创新产业，配以学校和周边住宅，形成湖岸周围风景秀丽的新综合体。

草海北岸被设计成为一个可以使区域内生态合理承受的可持续性社区范例，其中一个关键性的要素就是恢复湖里被污染的水体（这在当前中国是很大的一个部分）。这种恢复主要要依靠重新为污水处理企业选址，清疏现有的沉积，减少磷和氮的排放，建设新的湿地来吸纳社区内的雨水以及重新恢复当地植被和野生动物等战略来实施。

除了重建当地的生态系统和恢复湖水的整体质量，这个规划同时也配合了其他的环境原则。这个紧凑的社区规划实现了公共交通的高效连接，创建了步行街网络，同时又保护并丰富了露天公共场所。其中意义最重大的就是对大官公园的修葺。怀着对该公园800年历史的敬意，它的传统核心将被更新，新开放的空间用于各种娱乐和教育活动。增加在现有的该区域独特品质之外的是，社区内的建筑为传统的当地村庄赋予了当代的主题。建筑物都朝向湖面，利用湖面的微风达到自然通风的效果，慷慨的公园和广场也鼓励了更多的户外活动及应用。

Project name: Caohai North Shore Conceptual Master Plan
Award date: 2006
Location: Kunming, China
Site area: 520 hectares
Architect: Sasaki Associates, Inc.
Client: Shui On Land Ltd.
Completion date: 2005
Award name: McGraw-Hill Construction 1st Bi-Annual "Good Design Is Good Business" China Awards 2006, Best Planning Project

项目名称：草海北岸概念性总体规划
获奖时间：2006
项目位置：中国 昆明
规划面积：520公顷
规划设计：佐佐木设计事务所
业主：香港瑞安房地产发展有限公司
完成时间：2005
所获奖项：麦格劳-希尔公司《建筑实录》、《商业周刊》第一届"好设计创造好效益"中国奖项 2006最佳规划设计

1. Model
   模型

2. Detail Area
   细部

3. Urban Design
   城市规划图

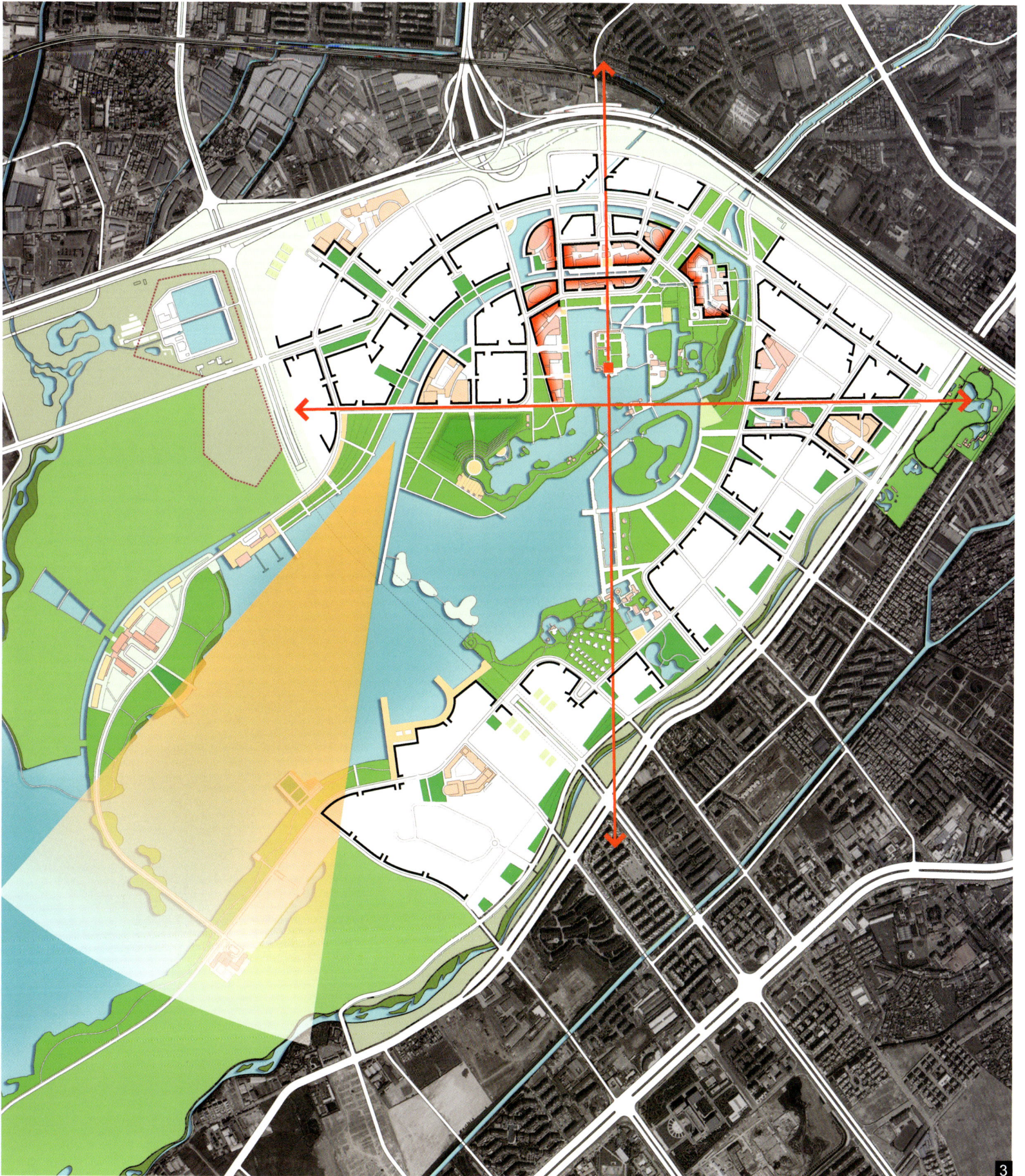

⟷ Axial 轴线	— Urban Edges 城市边界	▬ Retail Edges 零售边界	◁ Sleeping Beauty Mountain View 面向睡美人山的视野	▮ Public Buildings 公建
▮ Public Parks 公园	▮ Community Parks 社区公园	▮ Dianchi Ecological Sport Park 滇池生态运动公园	▮ Wetlands 湿地	

NORTH

1. Daguan Park Master Plan
   大观公园平面图

2. Perspective
   手绘效果图

3. Perspective
   手绘效果图

1. Daguan Perspective
   大观公园手绘效果图

2. Daguan Perspective
   大观公园手绘效果图

3. Daguan Perspective
   大观公园手绘效果图

Qiaonan Village Historic Preservation Scheme

泉州桥南古村再生规划设计

Quanzhou is a national-renown historic city. The Luoyang Bridge to the east of the site, as one of the cradles for the Silk Road, is scheduled to bid for World Heritage Site. Then, the Qiaonan Village, integrated with the Luoyang Bridge, should play what kind of a role? What should be most valued? How could we revitalize this somewhat declining old village? These questions become keys for the project.

The designers from AECOM Design + Planning followed a cross-discipline concept, and worked closely with experts from the fields of economy, planning, design, environment and ecology. They cooperated with local professionals, leaders and residents to produce a series of sustainable solutions: renovation of the old streets, reconstruction of preservation structure, connection to nature, return to ecology and beyond, the rebirth of regional culture, new definition of the site, economic vigour as a catalyst, and integration of function and culture.

The project won the Planning Award from "Architectural Record" in 2006. The ceremony was held during the first Global Construction Summit in China, and the project was on an exhibition tour afterwards. Every team member of the project feel much encouraged as they recall the process, being proud of their contribution to the preservation of the historic village and envisioning the cultural tour coming soon.

泉州是国家级历史文化名城，基地东侧的洛阳桥作为海上丝绸之路的起点之一，正在积极准备申报世界自然文化遗产，那么与洛阳桥融为一体的桥南村应该扮演什么样的角色？什么是其最具价值的？如何让略显没落的千年村落再现昔日的辉煌和活力？成为项目的关键。

AECOM规划＋设计的项目小组采用多学科整合的规划设计理念，经济、规划、设计、环境、生态专家携手工作，紧密和当地专家和领导以及原住民开展全方位协作，制定了一系列可持续发展解决途径：再现千年古街等城市遗产、建构新的保护框架、连接自然山水、生态文化的回归与嬗变、赋予文化的再生和永恒、场所的新定义、利用经济活力的催生、功能与文化的统一与永恒。

本项目在2006年荣获美国《建筑实录》规划设计奖，并在中国首届全球建筑峰会期间在中国颁奖和巡展，每一位参加过本项目的工作人员备受鼓舞，联想到本项目进行过程中的一幕一幕，此情此景，感到为中国的历史文化村落保护和即将开展的文化之旅所尽的微薄之力，不枉此行。

Project name: Qiaonan Village Historic Preservation Scheme
Award date: 2006
Location: Fujian, China
Site area: 51.27 hectares
Architect: AECOM
Client: Quanzhou Luojiang Real Estate Co.
Completion date: 2005
Award name: McGraw-Hill Construction 1st Bi-Annual "Good Design Is Good Business" China Awards 2006, Best Planning Project

项目名称：泉州桥南古村再生规划设计
获奖时间：2006
项目位置：中国 福建
规划面积：51.27公顷
规划设计：AECOM
业主：泉州洛江城市建设开发有限公司
完成时间：2005
所获奖项：麦格劳–希尔公司《建筑实录》、《商业周刊》第一届"好设计创造好效益"中国奖项 2006最佳规划设计

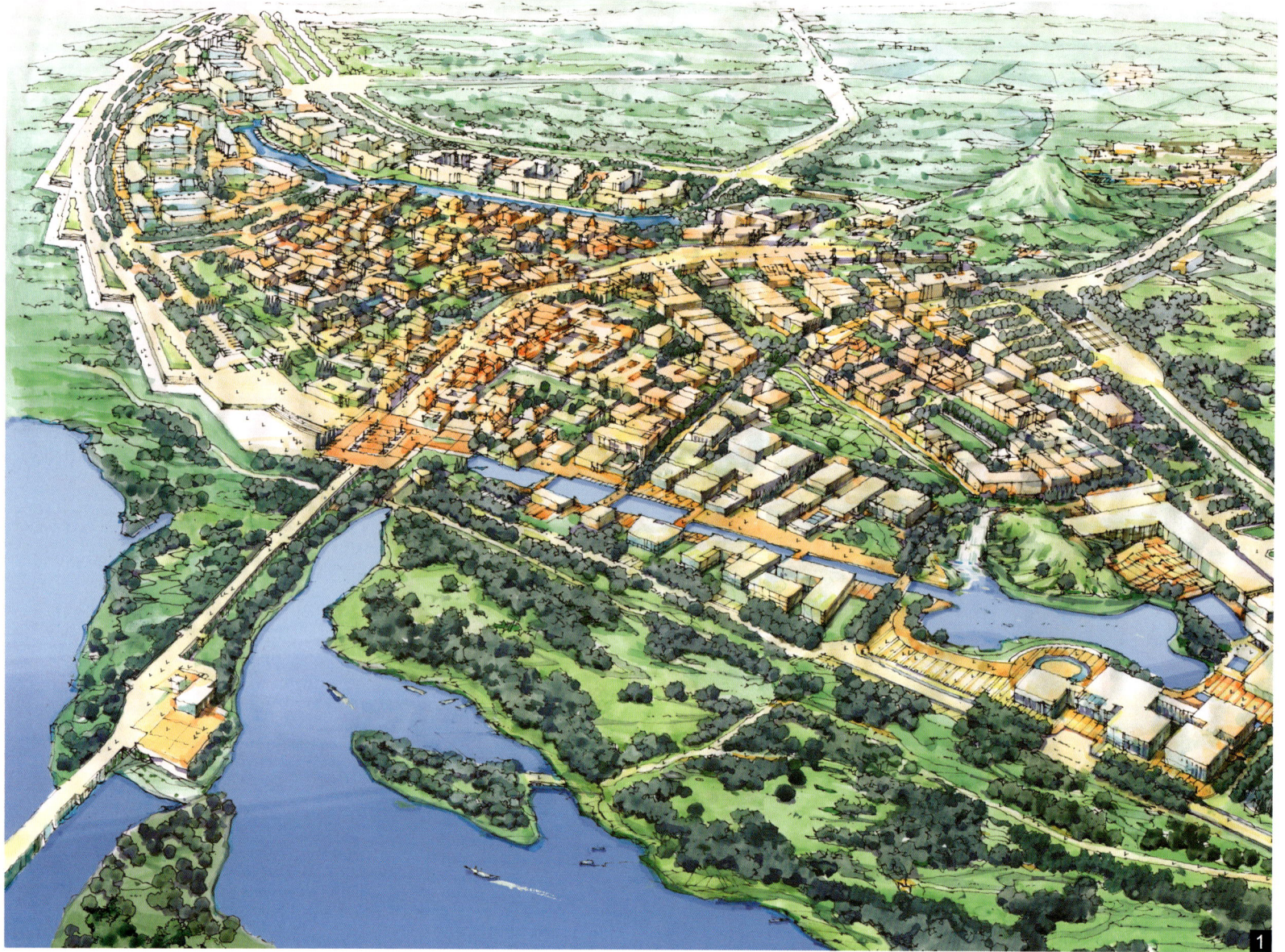

1. Bied View
   鸟瞰图

2. Haisi Square
   海丝广场

3. Cai Xiang Temple
   蔡襄祠

4. Countryard Park
   乡野公园

1

1. Waterfront Seafood Restaurant
   滨海餐馆
2. Museum of Bridges
   桥梁博物馆
3. Sea Silk Road Promenade
   海上丝绸之路步行道
4. Wetland Park with Boardwalk
   湿地公园
5. Luoyang Bridges History Museum and Visitor Center
   洛阳桥历史博物馆和游客中心
6. Stone Cutters Museum
   石雕博物馆
7. Oyster House
   牡蛎主题屋
8. Farmers Garden
   乡村公园
9. Stage of Traditional Folk
   南音戏台
10. Five Star Hotel
    五星级宾馆
11. Entry Plaza and Undergroun Parking
    入口广场与地下停车场
12. Sea Silk Plaza
    海丝广场
13. Villa Residential
    别墅式住宅
14. Clubhouse for Villa Residential
    别墅社区会所
15. Boutique Hotel
    家庭旅馆
16. Primary School
    小学
17. Bridge and Island
    桥与岛
18. Restaurant on Island
    岛上餐厅
19. Main Street
    古街
20. Renovation of Buildings along Main Street
    沿街进行更新
21. Liu Family Temple and Museum
    刘氏家庙与博物馆
22. Cai Xiang Temple Square
    蔡襄祠广场
23. Walking Tour
    滨水商业游线
24. Weekend Market / Night Market
    周末市场/夜市
25. Community Facilities
    社区设施
26. Public Neighborhood Park
    公共的社区公园
27. Buffer Trees
    红树林
28. Bus and Passenger Car Parking Lot
    公共汽车与旅游小汽车停车场

1. Plan
   总平面图
2. Square
   广场
3. View from Far Away
   远景

Horizontal Skyscraper – Vanke Center

万科中心

Hovering over a tropical garden, this "horizontal skyscraper" is a hybrid building including apartments, a hotel, and offices for the headquarters for Vanke Co. ltd. A conference center, spa and parking are located under the large green, tropical landscape which is characterized by mounds containing restaurants and a 500-seat auditorium.

The building appears as if it were once floating on a higher sea that has now subsided; leaving the structure propped up high on eight legs.

The decision to float one large structure right under the 35-meter height limit, instead of several smaller structures each catering to a specific program, generates the largest possible green space open to the public on the ground level. The underside of the floating structure becomes its main elevation – the sixth elevation – from which "Shenzhen Windows", offer 360-degree views over the lush tropical landscape below. A public path beginning at the "dragon's head" will connect through the hotel and the apartment zones up to the office wings.

As a tropical strategy, the building and the landscape integrate several new sustainable aspects. A micro-climate is created by cooling ponds fed by a greywater system. The building has a green roof with solar panels and useslocal materials such as bamboo. The glass facade of the building will be protected against the sun and wind by porous louvers. The building is a Tsunami-proof hovering architecture that creates a porous micro-climate of public open landscape.

围绕着一个热带花园，万科企业股份有限公司的总部大楼就是这样一座包括公寓、酒店、办公等综合功能的"躺着的摩天大楼"。会议中心、SPA理疗和停车厂都位于一个大型绿色的带有餐厅和500坐席礼堂的热带景观带之下。这座建筑就好像曾经在海里漂浮后来搁浅了一样，留下了八条腿支撑着的结构。

这个设计没有采用几个小部分后期结合的模式，而是直接选用了在35米高度限制下的整体大漂浮结构，在底层为公众保留了最大程度的绿色开放空间。在漂浮结构的下方就是建筑的主要海拔高度所在，"深圳之窗"的第六高海拔。在这里可以享受到下面360度的热带风景。一条公路从"龙头"开始蜿蜒，穿过酒店和公寓与公司的两翼相连。

为了适应热带气候，大厦和景观设计都考虑到了一些承受温度的方面。中水系统形成的凉水池为办公空间提空了宜人的小气候。大厦的屋顶是带有太阳能板的绿色结构，并采用了当地的材料，比如竹子。大厦正面的玻璃利用多孔的天窗来遮蔽阳光和风。整个建筑形成了内部空间的小气候，并且符合防海啸的建筑学要求。

Project name: Horizontal Skyscraper – Vanke Center
Award date: 2010
Location: Shenzhen, China
Building area: 120,445 m²
Architect: Steven Holl Architects, CCDI
Client: China Vanke Co., Ltd.
Photographer: Steven Holl Architects, Iwan Baan, Shu He
Completion date: 2009
Award name: McGraw-Hill Construction 3rd Bi-Annual "Good Design Is Good Business" China Awards 2010, Best Green Project

项目名称：万科中心
获奖时间：2010
项目位置：中国 深圳
建筑面积：120,445平方米
建筑设计：斯蒂文霍尔建筑事务所   中建国际
业主：万科企业股份有限公司
完成时间：2009
所获奖项：麦格劳–希尔公司《建筑实录》、《商业周刊》第三届"好设计创造好效益"中国奖项   2010最佳绿色设计

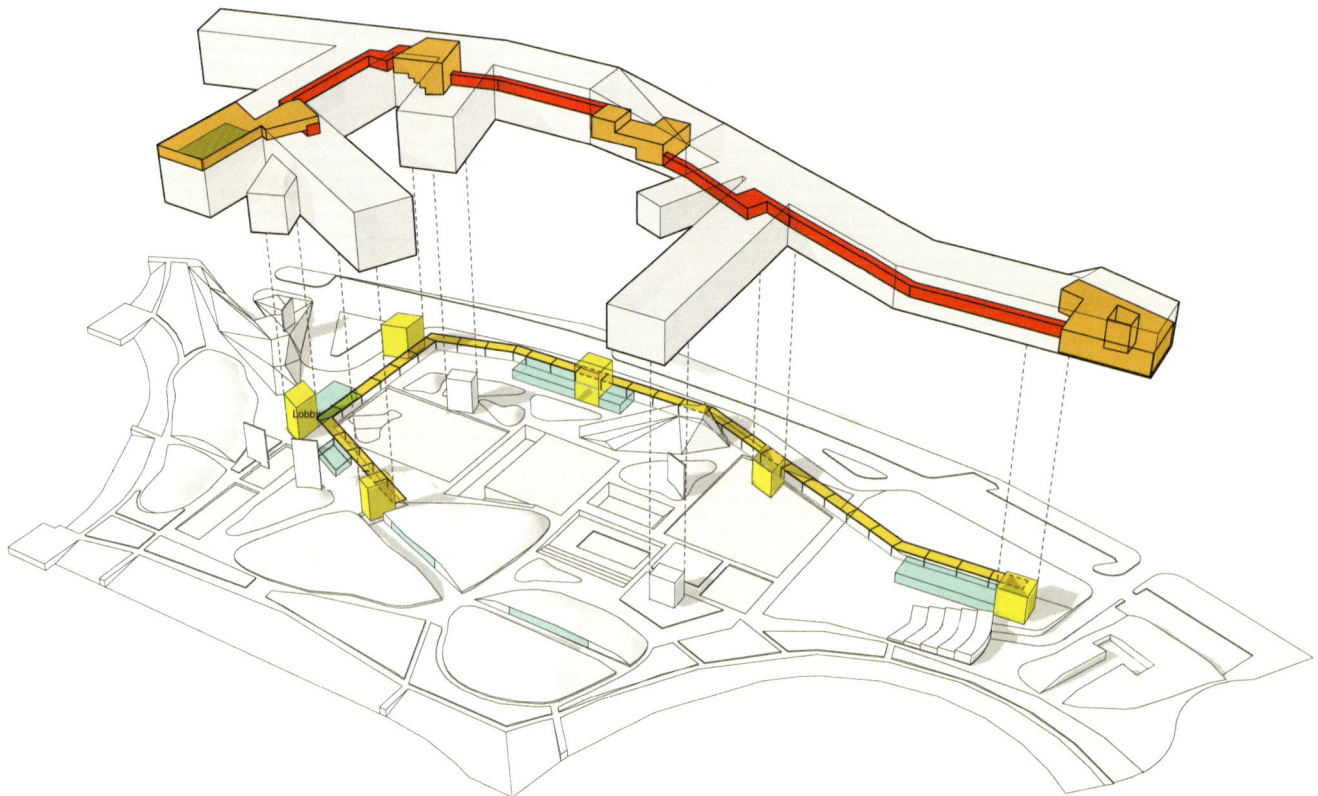

1

2

1. Water under the Building
   大楼下的水景
2. Full View
   全景
3. Night View
   夜景
4. Pool
   水池

1

2

1. Interior
   室内
2. Stairs
   楼梯
3. Interior
   室内
4. Elevator
   电梯
5. Structure
   结构图

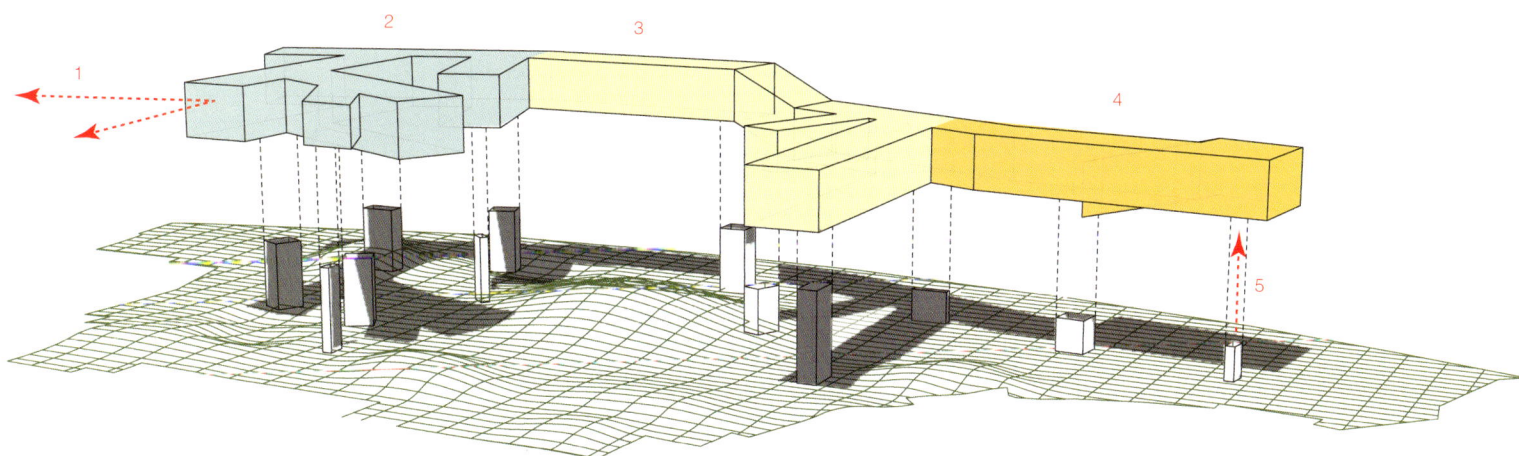

1. Ocean Views
   海景
2. Office
   办公室
3. Apartments
   公寓
4. Hotel
   酒店
5. Stairs+Elevators
   楼梯和电梯

IBR Headquarters

建科大楼

Situation

IBR Headquarters is situated on a site with an area of 3,000 square meters. The total construction area is 18,200 square metres, including twelve floors above ground and two floors underground, with a height of 57.9 meters and a plot ratio of 4.0. The tower comprises labs, offices, meeting rooms, underground parking lots, restaurants, and apartments, etc.

Technologies

Low budget, soft tech, and passive mode are the essence of the more than forty applied technologies, in which passive, low-costing and managerial techniques take up 68%. The following aspects fully demonstrate this essence: ventilation, lighting, greenery, and waterscape are explored to create outdoor public spaces that connect human and nature; the combination of active and passive techniques such as natural ventilation and lighting, efficient air conditioning, lighting facilities, simple and human-centered interior design; energy-saving, recycling resources, water-economy, material-efficiency are all factors employed to reach the goal of saving resources and efficient running. In the whole life cycle of the building, maximal energy-saving and resource-utilization solutions would help protecting environment and reducing pollution. With 2/3 of materials needed for local constructions of similar scale, the project received level 3 Green Architecture Standard of China (the highest level in China).

Results

The survey on the functions of IBR Headquarters shows that compared with local buildings of the same type, energy consumption of air conditioning reduced 50% per square meter per year; energy consumption of lighting reduced 71%; total energy consumption reduced 59%. At the same time, high quality offices are provided; 87% of the occupants expressed their satisfaction towards their working environment. Working efficiency has been remarkably improved.

Social benefits

Every year, 610 tons of coal could be saved, together with 1,450,000 RMB saved due to power-efficiency and 54,000 RMB due to water-efficiency, equaling a reduction of 1,622 tons of CO_2 emission. As a base for exhibiting construction technologies and arts, and also for the popularization of green architecture technologies, it receives 8,000 visitors annually.

概况

建科大楼用地面积3000平方米，总建筑面积1.82万平方米，地上12层，地下两层，总建筑高度57.9米，容积率4.0。主要功能包括实验、办公、会议交流、地下停车、休闲餐饮、专家公寓等。

技术体系

建构以低成本、软技术、被动式为核心的技术体系，共采用40多项绿色建筑技术（其中被动、低成本和管理技术占68%）。突出体现在以下几个方面：充分利用风、光、绿化、水景等元素营造了人与自然共享的室外活动空间；整合运用通风、采光、高性能空调与照明设施、简约和人性化室内装修等主被动结合技术营造了健康室内空间；系统采用了各类节能、可再生能源、节水、节材等技术实现资源节约和高效运营。在建筑全寿命周期内最大限度节约和高效利用资源、保护环境、减少污染，以当地同类建筑2/3的建安成本达到绿色建筑三星级（中国国家标准最高等级）的要求。

实际运行效果

根据建科大楼实际运行能耗监测数据，与当地同类建筑平均水平比较，年单位建筑面积空调能耗低50%，照明能耗低71%，总能耗低59%。同时提供高品质的办公环境，使用者对室内环境的满意率为87%，工作效率显著提升。

社会效益

每年可节约电费145万元、水费5.4万元、标煤610吨，相当于减排二氧化碳1622吨。作为定期向市民开放的建筑技术、艺术展示基地及绿色建筑技术科普基地，每年到访参观交流人数逾8000人次。

Project name: IBR Headquarters
Award date: 2010
Location: Shenzhen, China
Building area: 18,200 m²
Architect: Shenzhen Institute of Building Research
Client: Shenzhen Institute of Building Research Co., Ltd.
Completion date: 2009
Award name: McGraw-Hill Construction 3nd Bi-Annual "Good Design Is Good Business" China Awards 2010, Best Green Project

项目名称：建科大楼
获奖时间：2010
项目位置：中国 深圳
建筑面积：18,200平方米
建筑设计：深圳市建筑科学研究院有限公司
业主：深圳市建筑科学研究院有限公司
完成时间：2009
所获奖项：麦格劳－希尔公司《建筑实录》、《商业周刊》第三届"好设计创造好效益"中国奖项 2010最佳绿色设计

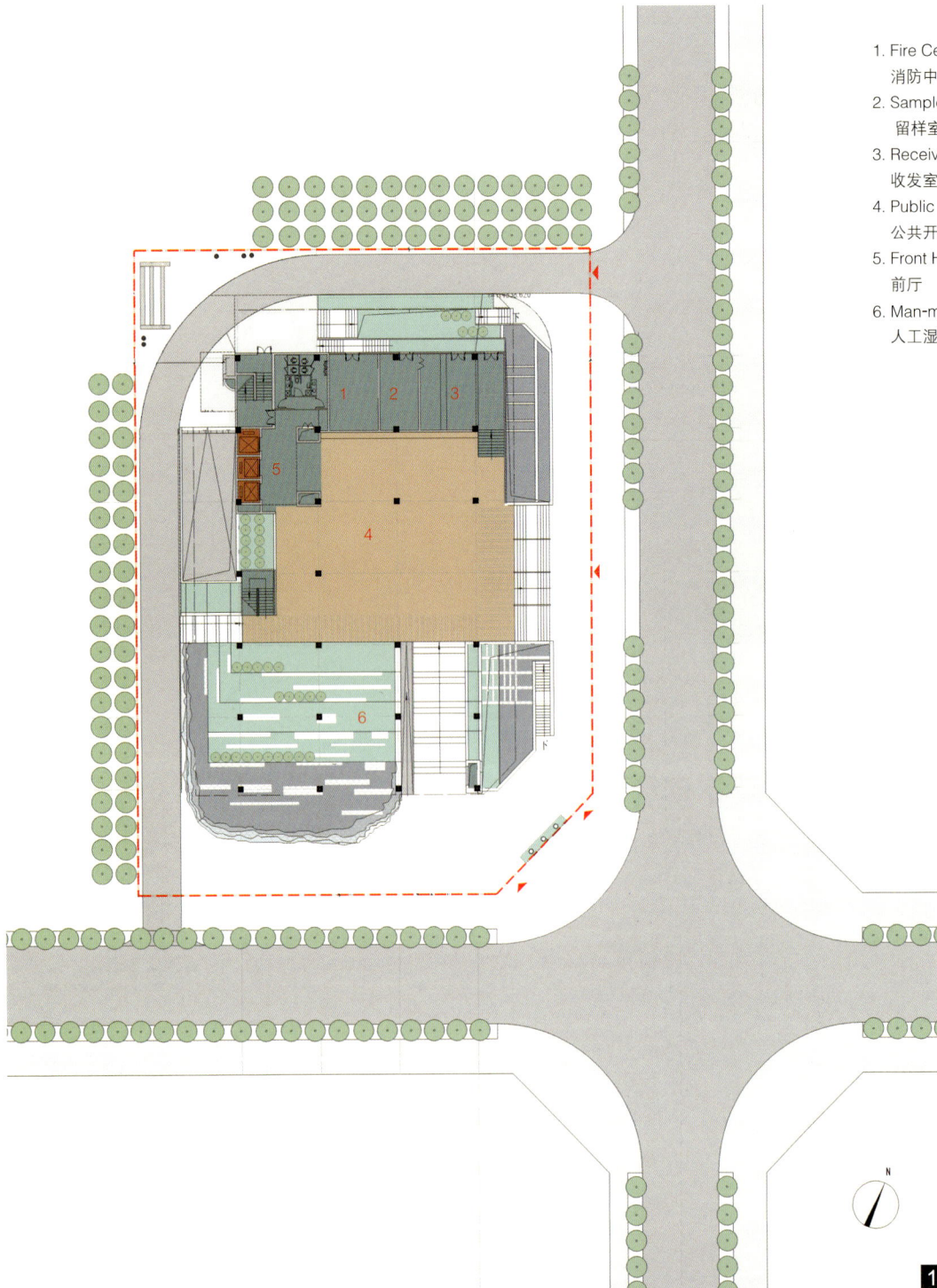

1. Fire Center and Intelligent Administration Room
   消防中心智能监控室
2. Samples Room
   留样室
3. Receiving
   收发室
4. Public Area
   公共开放区
5. Front Hall
   前厅
6. Man-made Wetland
   人工湿地

1. First Floor Plan
   一层平面图
2. Solar and Wind Power on the Roof
   屋顶太阳能、风力发电
3. Detail of Exterior
   外立面细部
4. Terrace for Relax
   休闲平台
5. Gardens on 6th floor
   6楼空中花园

1. Green on the Roof
   屋顶绿化

2. Naturally Lighted Water Pool ar East Entrance
   东入口处的玻璃水池自然采光

3. Fountain at the Entrance
   入口水景喷泉

4. Green Terrace
   绿化平台

1. Offices without Daytime Lighting
   白天不用电灯照明的办公空间

2. Naturally Lighted and Ventilated Lecture Hall
   可自然采光和通风的报告厅（开启状态）

3. Lobby of Lecture Hall on the Fifth Floor
   五层报告厅的前厅

4. Open Great Hall
   开放式大堂

Chongqi Channel Environmental Landscape Plan

崇启通道生态景观规划

As one of the most important strategic redevelopment projects in Shanghai, Chongming Island enjoys the best ecological environment in Shanghai, being named the "last eternal land in Shanghai". The long history of enclosure farming has created the unique linear streamways, villages and the pattern of the farms. Meanwhile, the eastern beach and the northern lake are ecological preservations that constitute distinctive ecological landscape for this area.

Chongqi project comprises two bridges, a channel that crosses the Yangtze River and a highway of thirty-two kilometers. The construction of this project would open a gate for Shanghai, leading to the north of the Yangtze River and thus making an end for the isolation history of Chongming Island. In this sense, it would be an epoch-making project. However, while the long highway brings opportunities for the island, it also endangers its integrity: the distinctive elements are to be separated and the special layout might be destroyed. Therefore, the key of the plan should be finding a solution to avoid – at least to reduce – the destruction of the landscape on the island while seeking to explore and integrate its ecology, landscape, tourism, culture, etc. in order to bring benefit to the island.

作为21世纪上海发展的战略重点之一，崇明岛是上海生态环境最好的地区，被誉为"上海最后的净土"。崇明岛的长期围耕历史使其形成了富有特色的线型河道，村落及农场肌理，同时岛上的东滩及北湖等生态保护区域，构成了特有的景观生态格局。崇启通道包括2座立交桥、穿越长江的隧道以及崇明岛上32公里长的高速公路，它的建设将为上海打开通向长江以北的大门，从而结束崇明岛长期与世隔绝的历史。从这个意义上来讲，这一项目无疑是划时代的。但是32公里长的线形高速通道在赋予崇明岛巨大机遇的同时，却也构成了对崇明各特色要素系统产生分割，破坏固有特色布局的危险。因此，规划设计的重点在于如何减少高速公路对崇明岛独特景观元素的破坏，同时，发掘并整合其生态、景观、旅游、文化等多元价值的机遇。

Project name: Chongqi Channel Environmental Landscape Plan
Award date: 2008
Location: Shanghai, China
Site area: 34.5 km²
Architect: AECOM
Client: Shanghai Municipal City Planning Administration
Completion date: 2007
Award name: McGraw-Hill Construction 2nd Bi-Annual "Good Design Is Good Business" China Awards 2008, Best Green Project

项目名称：崇启通道生态景观规划
获奖时间：2008
项目位置：中国 上海
规划面积：34.5平方公里
规划设计：AECOM
业主：上海市城市规划管理局
完成时间：2007
所获奖项：麦格劳－希尔公司《建筑实录》、《商业周刊》第二届"好设计创造好效益"中国奖项　2008最佳绿色设计

1. Rendering of the Entry
   入口效果示意

2. Site Plan
   区位

3. Master Plan and Visual Experience
   总图及视觉体验

4. Tourism
   休闲旅游

5. Habitat Connection Strategy
   栖息地连接策略

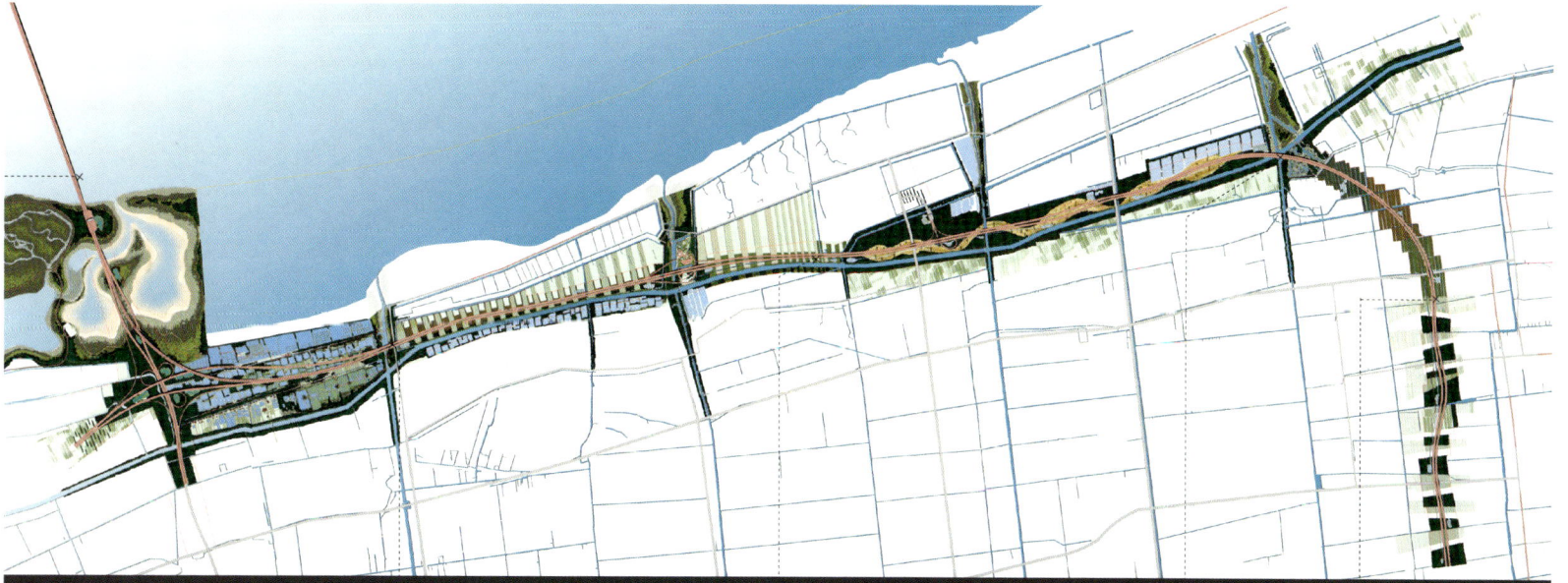

3

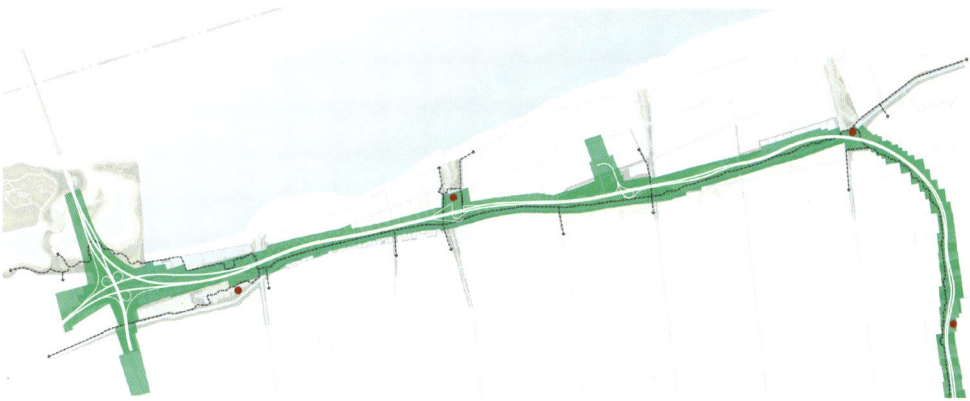

4

Highway 高速公路用地
Pedestrian System 步行系统
Main Scenic Point 主要景点

5

1. Primary Wildlife Corridor
   主要野生通道
2. Riparian Corridor
   河岸边通道
3. Secondary Wildlife Corridor
   次要野生通道
4. Primary Wetland Habitat
   主要动物栖息地
5. Storm Water Wetland Habitat
   自然暴雨湿地
6. Animal Crossing
   动物交叉口

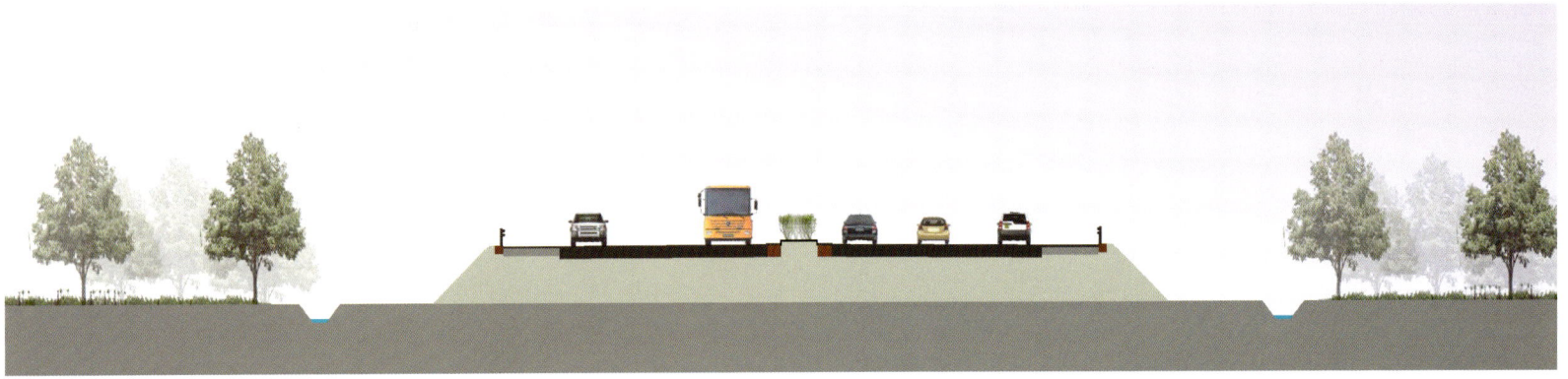

Current Section 现有断面

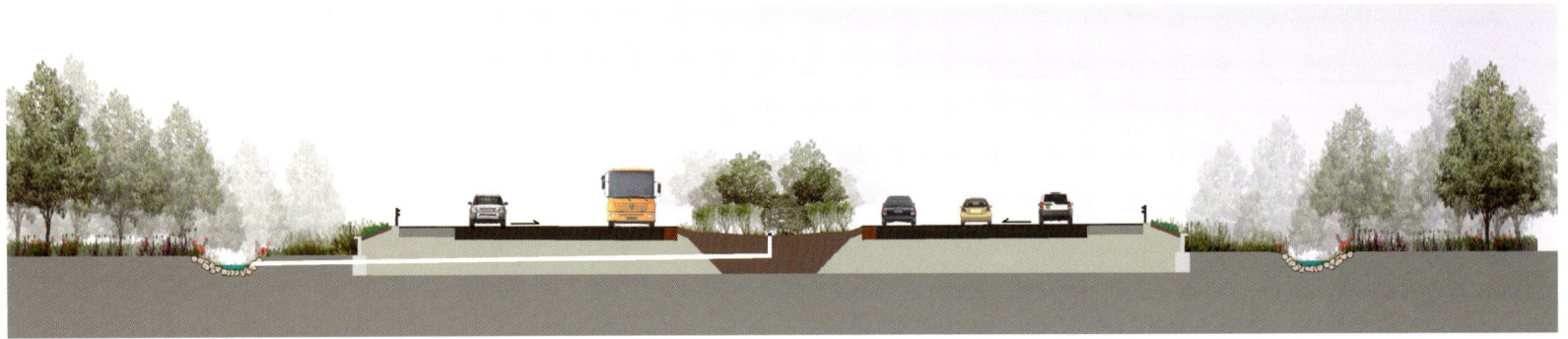

Proposed Section 设计断面

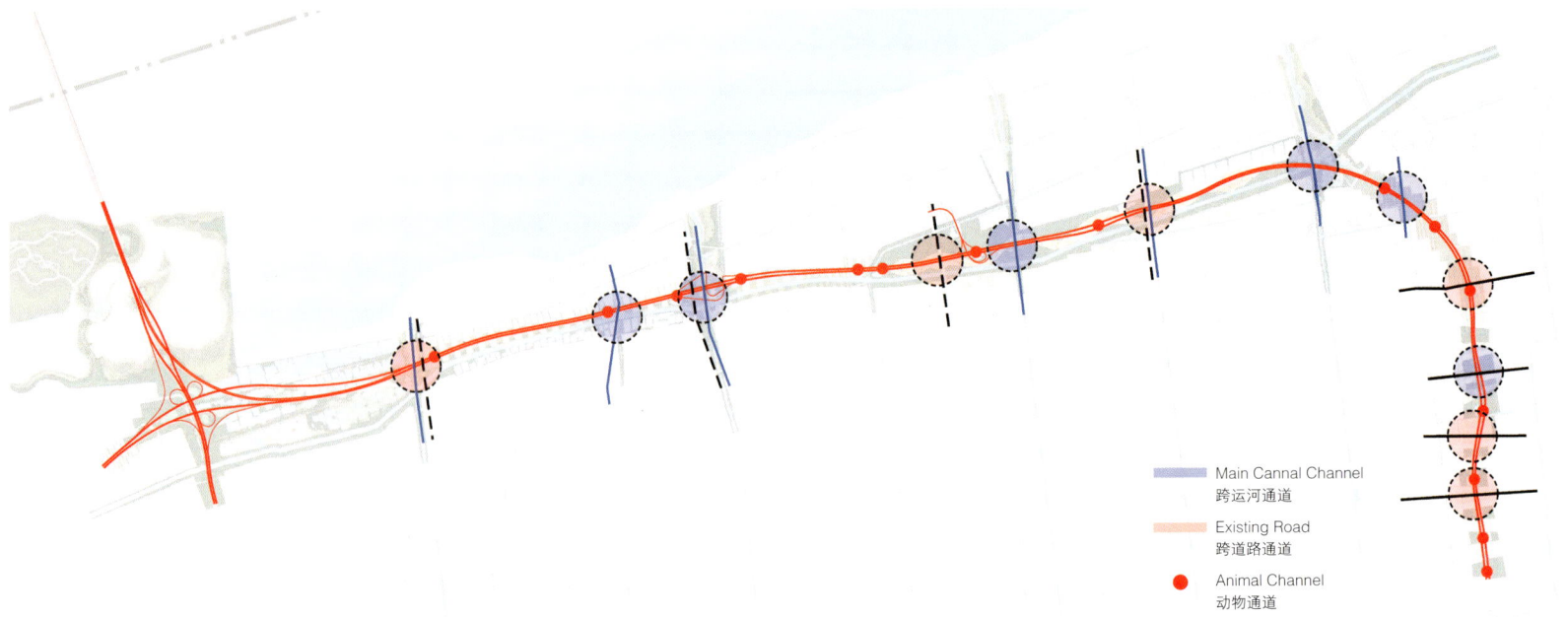

Main Cannal Channel
跨运河通道

Existing Road
跨道路通道

● Animal Channel
动物通道

2

Storm Water Wetland Systems
雨水处理湿地

● Planned Road
路面雨水收集点

3

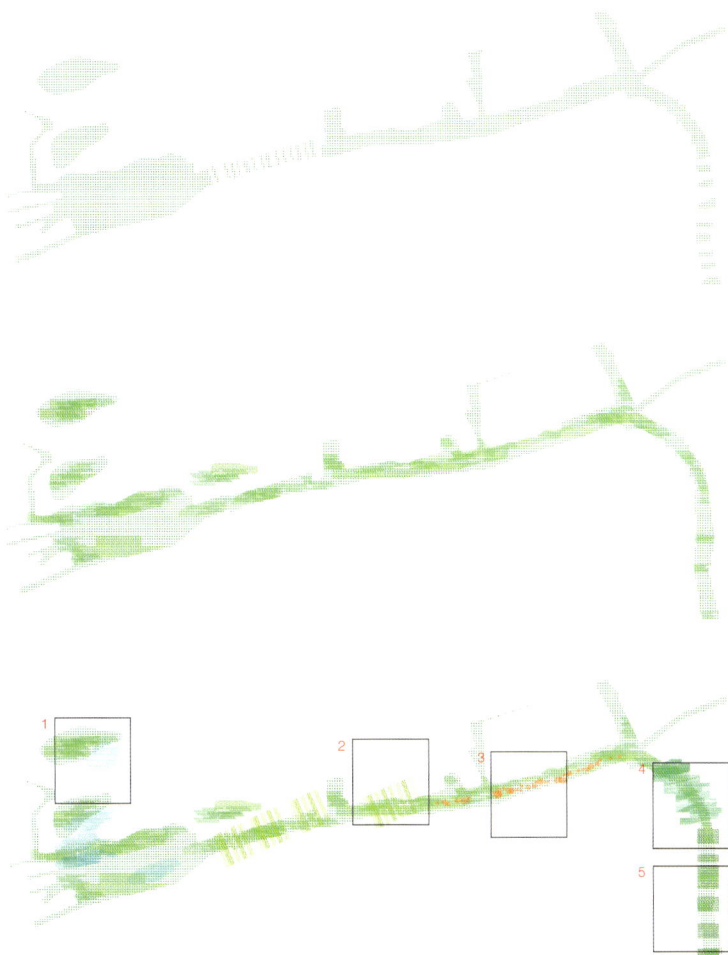

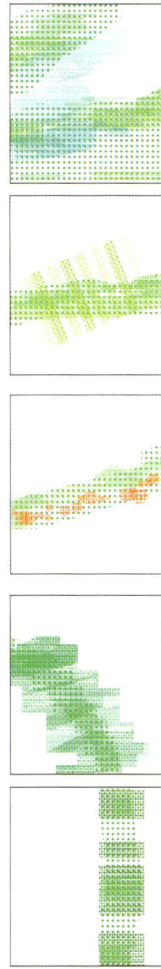

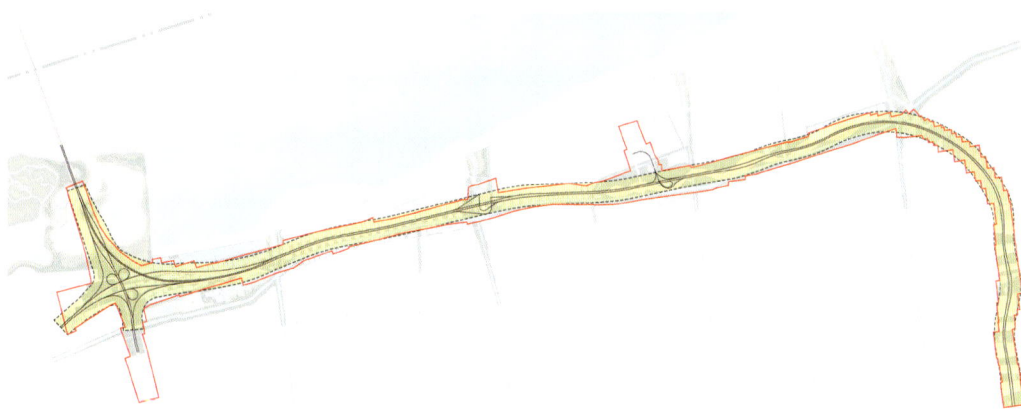

Legend:

— Proposed Buffering Zone
建议防护绿带

- - - Existing Buffering Zone
原有防护绿带

— Road Center Line
道路中心线

1. The Revise of Section Drawing
   断面调整

2. Main Channel
   通道

3. Rain Treatment
   雨水处理

4. Plants
   植栽

5. Protective Line
   防护带

Chongming North Lake District Master Plan

崇明北湖区总体规划

Jiangsu Province and the city government of Shanghai co-initiated the Chongqi Highway project in 2006, connecting Shanghai to the north, and putting the once isolated Chongming Island in a strategic position. AECOM was invited to participate in the Chongming North Lake District Master Plan competition at the end of 2006. The competition was held to create 34.5-square-kilometer holiday resorts in the northern part of Chongming Island.

The North Lake District occupies a 7.7-hectare area situated at the Yangtze River estuary. It contains the largest salt lake in East China and an important habitat for birds. Thus issues like environmental-friendly development and preservation became the primary concern. Consequently, the master plan started with protecting the environment. A natural recycling system was designed for the preservation of ecologically safe water. Meanwhile, special researches on the eco-system of animals and plants were conducted to establish a regulation for the habitat preservation. Based on such preparations, well-controlled resort projects could begin. The sizes of these resorts should not be too large. The essence of the projects should not be their sizes; rather, it lies in the integration of different resorts and the balance between tourism and ecological environment. In addition, the master plan proposed a series of landscape solutions to adapt to the particular saline-alkali soil in the North Lake District.

2006年，江苏省和上海市政府共同启动了崇启通道高速公路项目，打开上海向北的门户，同时也将曾经与世隔绝的崇明岛摆在了连接大区域的战略位置上。2006年底，AECOM公司接到邀请，参加上海市规划局崇明北湖区域总体规划国际设计竞赛：在崇明岛北部打造34.5平方公里的生态旅游度假区。

面对北湖这个7.7公顷，位于长江入海口的华东地区最大的咸水湖和重要的鸟类栖息地，如何平衡开发和保护变得尤为敏感。规划首先从环境保护入手，为北湖设计自然循环系统，以保证水体水质生态安全，同时研究区域动植物生态系统，确立完整栖息地保护框架，在此基础上谨慎的引入旅游度假项目，控制开发的尺度，通过不同游线和活动的组织，整合区域旅游资源，使小尺度的开发发挥最大的效能，从而使旅游开发与生态环境相兼容。此外，规划还提出一系列植被景观策略，以适应北湖区域特有的盐碱土壤环境。

Project name: Chongming North Lake District Master Plan
Award date: 2008
Location: Shanghai, China
Site area: 34.5 km²
Architect: AECOM
Client: Shanghai Municipal City Planning Administration
Completion date: 2007
Award Name: McGraw-Hill Construction 2nd Bi-Annual "Good Design Is Good Business" China Awards 2008, Best Green Project

项目名称：崇明北湖区总体规划
获奖时间：2008
项目位置：中国 上海
规划面积：34.5平方公里
规划设计：AECOM
业主：上海市城市规划管理局
完成时间：2007
所获奖项：麦格劳–希尔公司《建筑实录》、《商业周刊》第二届"好设计创造好效益"中国奖项　2008最佳绿色设计

1. Main Bird View
   总体鸟瞰图

2. Bird View of North Lake Village
   北湖村鸟瞰图

3. Bird View of North Lake Pier
   北湖游艇码头鸟瞰图

4. Bird View of the Residential Area on the Island
   岛屿住宅区鸟瞰图

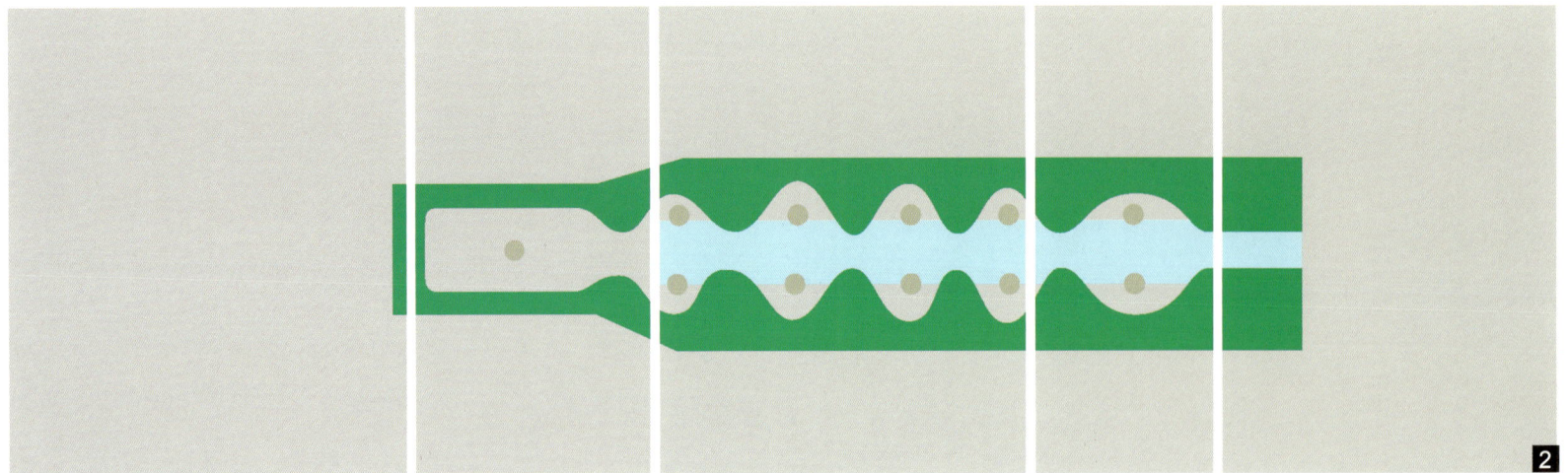

1. Plan
   总平面图

2. Ecology Plants Arrangement
   生态植被分区

3. Site Plan
   区位

1

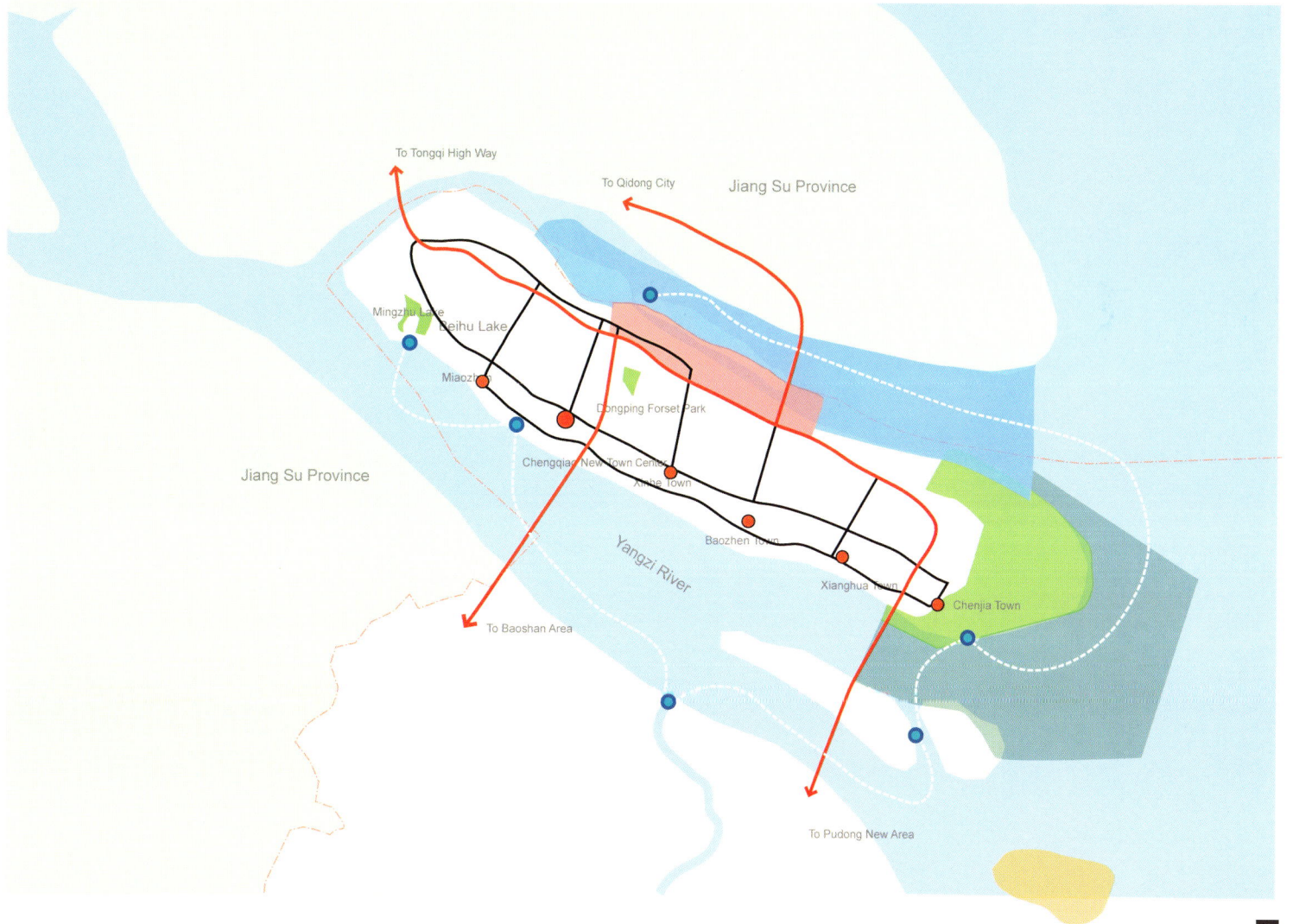

To Tongqi High Way

To Qidong City

Jiang Su Province

Mingzhu Lake

Beihu Lake

Miaozhen

Dongping Forset Park

Chengqiao New Town Center

Xinhe Town

Baozhen Town

Xianghua Town

Chenjia Town

Jiang Su Province

Yangzi River

To Baoshan Area

To Pudong New Area

3

Sunny Bay Station

迪士尼线欣澳站

Sunny Bay Station is 200 m long and 40 m wide and was constructed around four existing train tracks. The station was planned with a minimum built area.

Arup identified sustainability as a driver in the design of Sunny Bay Station, one of the greenest stations operating in Hong Kong. Natural ventilation is used in the public areas of the station, replacing energy-hungry air-conditioning.

The energy-efficient bioclimatic roof reduces energy consumption by 70%, compared with a traditional cooling system. Made of polytetrafluoroethylene (PTFE) fabric, the aesthetics of the roof reflect the leisurely nature of Sunny Bay Station. The PTFE fabric is translucent and self-cleaning, reducing both the need for artificial lighting and future maintenance costs.

Computational fluid dynamics (CFD) studies were carried out during the design to establish the concept and confirm the thermal comfort criteria. One key concept was to infill the trusses with glazing, affecting a light and airy appearance while injecting a rhythm into the roof shape, which can be lost in surfaces made entirely of fabric. The glazed sections also provide access to the upper surface of the roof and incorporate smoke discharge vents. Low height Automatic Platform Gates are installed on the platform edges to increase passenger safety without inhibiting the flow of natural ventilation and a fine water mist system also cools passengers on Hong Kong's hottest days.

As this station provides the interconnection to the Disneyland Resort Line a different passenger profile uses this station – holiday makers and families, rather than daily commuters as at other MTR stations. The design team therefore developed a design that injects a note of whimsy and fun without compromising the function, robustness, or maintainability of structural elements.

The station includes two cross-track footbridges, ramps, escalators and elevators to cater to luggage-toting holiday makers and families with strollers, the elderly or mobility impaired passengers. Most passengers coming from the direction of central Hong Kong interchange to the themed Disneyland Resort Line trains by walking across the platform without changing levels.

欣澳地铁站长200米，宽40米，位于原有4条路轨旁。该站大堂是香港所有车站之中最小的一个。该站以环保为设计理念，结合了可持续性的设计元素，旨在打造香港绿色运输通道。公共空间中采用自然通风，尽可能地降低能源消耗，从而减少经营成本。

屋顶的设计根据生物气候学原理，采用节能处理，与传统的制冷系统相比，该设计可减少能源消耗达30%。聚四氟乙烯材料令屋顶更为美观的同时，彰显该站怡然、舒适的特点。透明的聚四氟乙烯材料，具有自动清洗功能，减少人工照明的需要，从而大大节省维修费用。

精良的高科技系统能够进行空气动力分析研究，以确保站内空气清新，环境舒适。钢桁架上安置的玻璃窗令内部空间通透、明亮，与顶棚独特的造型相得益彰，并与排烟孔相通。低矮的自动月台闸门安装在平台的两边，确保乘客安全，便于自然通风。

该站作为香港迪士尼乐园的大门，每天接待的乘客身份也与普通的地铁站不同，乘客涉及度假人士和家庭等等。设计师通过巧妙设计，在确保空间功能性的同时，为其注入了强烈的戏剧色彩及浓厚的神秘气氛。

车站中设有两个跨轨道天桥、坡道、自动扶梯和电梯，为负重的乘客、老人及残障人士提供便利条件。来自香港市中心的乘客在此转站，只需跨过同层的站台，即可换乘迪士尼线路。

Project name: Sunny Bay Station
Award date: 2008
Location: Hong Kong, China
Project area: 8,000 m²
Architect: Arup and Aedas Limited
Client: MTR Corporation
Photographer: Arup
Completion date: 2009
Award name: McGraw-Hill Construction 2nd Bi-Annual "Good Design Is Good Business" China Awards 2008, Best Green Project

项目名称：迪士尼线欣澳站
获奖时间：2008
项目位置：中国 香港
项目面积：8,000平方米
建筑设计：奥雅纳工程顾问 凯达环球有限公司
业主：香港铁路有限公司
摄影师：奥雅纳工程顾问
完成时间：2009
奖项名称：麦格劳–希尔公司《建筑实录》、《商业周刊》第二届"好设计创造好效益"中国奖项 2008最佳绿色设计

1. Full View of the Station
   站台远景

2. Plan
   平面图

3. Appearance
   外景

4. Bird View
   鸟瞰

5. Entrance
   出入口

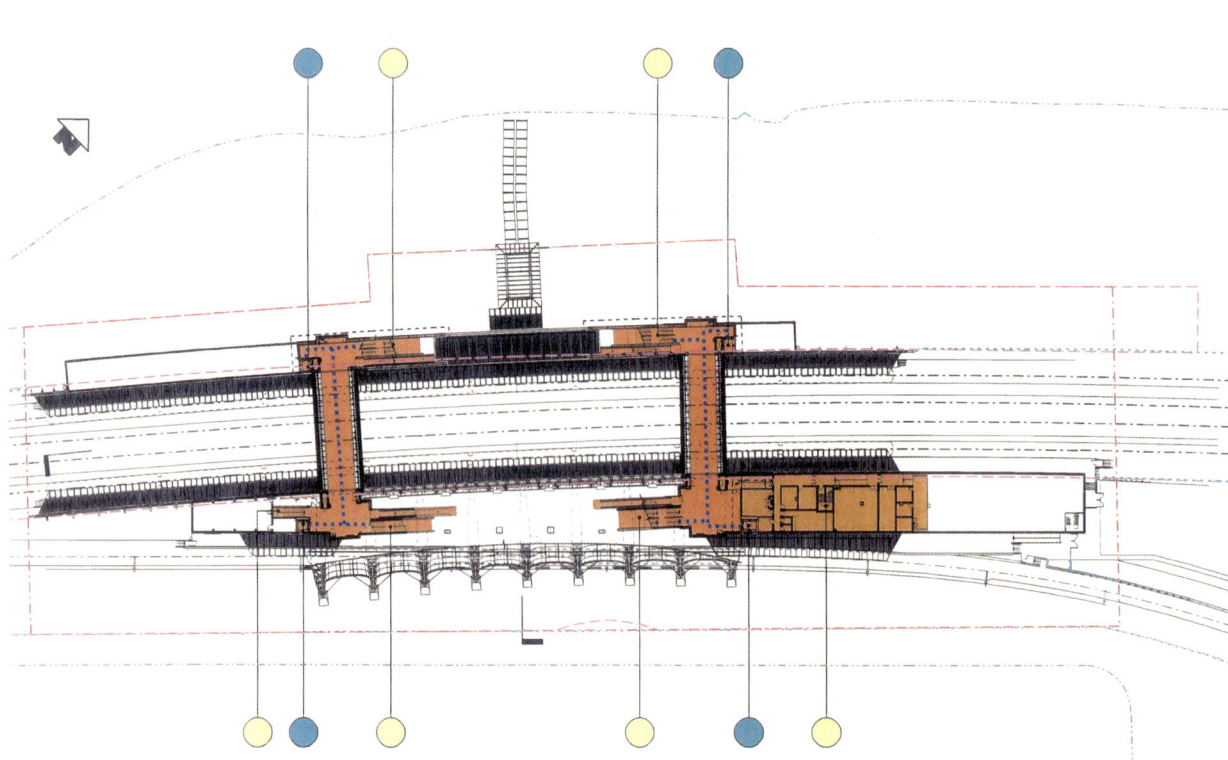

1. Principal Fireman's
   Access Point
   消防通道

2. Means of Escape Stair
   紧急逃生口

3. Passenger Lifts
   乘客电梯

4. Emergency Access
   紧急通道

2

1. Area for Getting on and off
   站台

2. Full View
   车站全景

3. Area for Getting on and off
   站台

4. Waiting Area
   等候区

2008 Beijing Olympic Green and Forest Park

2008北京奥林匹克公园

In 2002, Sasaki was awarded first prize for its planning and urban design of the Olympic Green, the principal venue of the 2008 Beijing Olympics. Sasaki's proposal is deeply connected to an environmental ideal that has its beginnings in the myth and legends of ancient China. The goal was to link ancient China to the present while recognizing the contemporary imperative of sustainable development. The design is further informed by Beijing's urban history, and draws inspiration from the great urban axes of the world. It has three fundamental elements:

The Forest Park. This land encompasses the area north of the central area of the Olympic Green. It is conceived as an ideal paradise from which Chinese civilization emerged millennia ago. The park is a sculpted landform of hills, forests, and meadows. Existing bodies of water are reformed into a "Dragon Lake". The pastoral nature of the Forest Park gives way to a more ordered spatial idea by using water to link the central area and the Asian Games beyond.

The Cultural Axis Beijing was founded on the basis of its north/south axis. The concept plan extends the axis some five kilometers through the Olympic Green site. The scale of the axis is monumental in order to emphasize its significance, yet it concludes with the serene simplicity of the Forest Park hills.

The Olympic Axis Set against the Cultural Axis at an acute angle, the Olympic Axis begins at the existing Asian Games stadium. It extends northwest, through the National Stadium by Herzog & de Meuron. This axis then continues to a Sports Heroes Garden, intersecting the Cultural Axis at Zhou Dynasty Plaza, which commemorates the Chinese contributions to city building. The axis terminates at the Memorial of Olympic Spirit.

The urban design plan for the Olympic Green is conceived as a framework for the long-term evolution of the district. Streets and pedestrian routes extend from the adjacent districts seamlessly through the site, while public transit stations connect the area to the larger Beijing transit system. Mixed-use development sites are identified for the post-Olympic era so that the district may become a dynamic yet natural extension of 21st-century Beijing.

2002年，佐佐木赢得了2008年北京奥林匹克运动会绿地与森林公园邀请赛。因为佐佐木的提案深切的契合了古老中国神话与传奇中所渲染的环境理想。这个提案目标是将古老中国与现代理念连接在一起，符合当今世界可持续发展的主题。设计结合了北京的古城历史，并从世界伟大的都市实践中得到了启发。设计有三个基本元素：

森林公园。这片土地包括奥运绿地中心地区的北部地区。这个设想作为千年前中国文明涌现的一个理想天堂。公园的地形经过雕琢出现了小山、森林和草甸。现有的水体被整合为"龙湖"，森林公园里田园牧歌式的自然风光，通过一脉水流将中心区域和亚运村相连，呈现出美妙的空间感觉。

北京文化轴线在南北历史轴线的基础上形成了。设计的概念将这条轴线在奥运绿地上扩展了5公里。轴线的巨大规模是为了凸显它的重大意义，一直延伸到森林公园内静谧的小山处。

奥林匹克轴从现有的亚运村开始，与文化轴形成一个锐角，并穿过德梅隆设计的国家体育场向西北延伸。这条轴线还经过运动英雄花园，与文化轴在周朝广场交汇，来纪念中国建筑对城市发展的贡献，最后终止于奥运精神纪念馆。

奥林匹克绿色城市规划被作为该区长期发展的一个框架。街道和步行线路连通了毗邻的街区，更多的公共交通站点也形成了更大的北京交通系统。多功能的建筑成为了奥林匹克时代的标志，并且使北京成为21世纪真正的城市典范并持续发展。

Project name: 2008 Beijing Olympic Green and Forest Park
Award date: 2006
Location: Beijing, China
Site area: 2,800 acres
Architect: Sasaki Associates, Inc
Associate planner and landscape architect: Tsinghua Planning Institute
Client: Beijing Municipal Commission of Urban Planning
Design date: 2002
Award name: McGraw-Hill Construction 1st Bi-Annual "Good Design Is Good Business" China Awards 2006, Best Green Project

项目名称：2008北京奥林匹克公园
获奖时间：2006
项目位置：中国 北京
规划面积：2,800英亩
规划设计：佐佐木设计事务所
合作规划及景观设计：清华规划学院
业主：北京市规划委员会
设计时间：2002
所获奖项：麦格劳–希尔公司《建筑实录》、《商业周刊》第一届"好设计创造好效益"中国奖项 2006最佳绿色设计

2

1. Model View Looking North
模型

2. Grading Concept for Forest Park
森林公园概念图

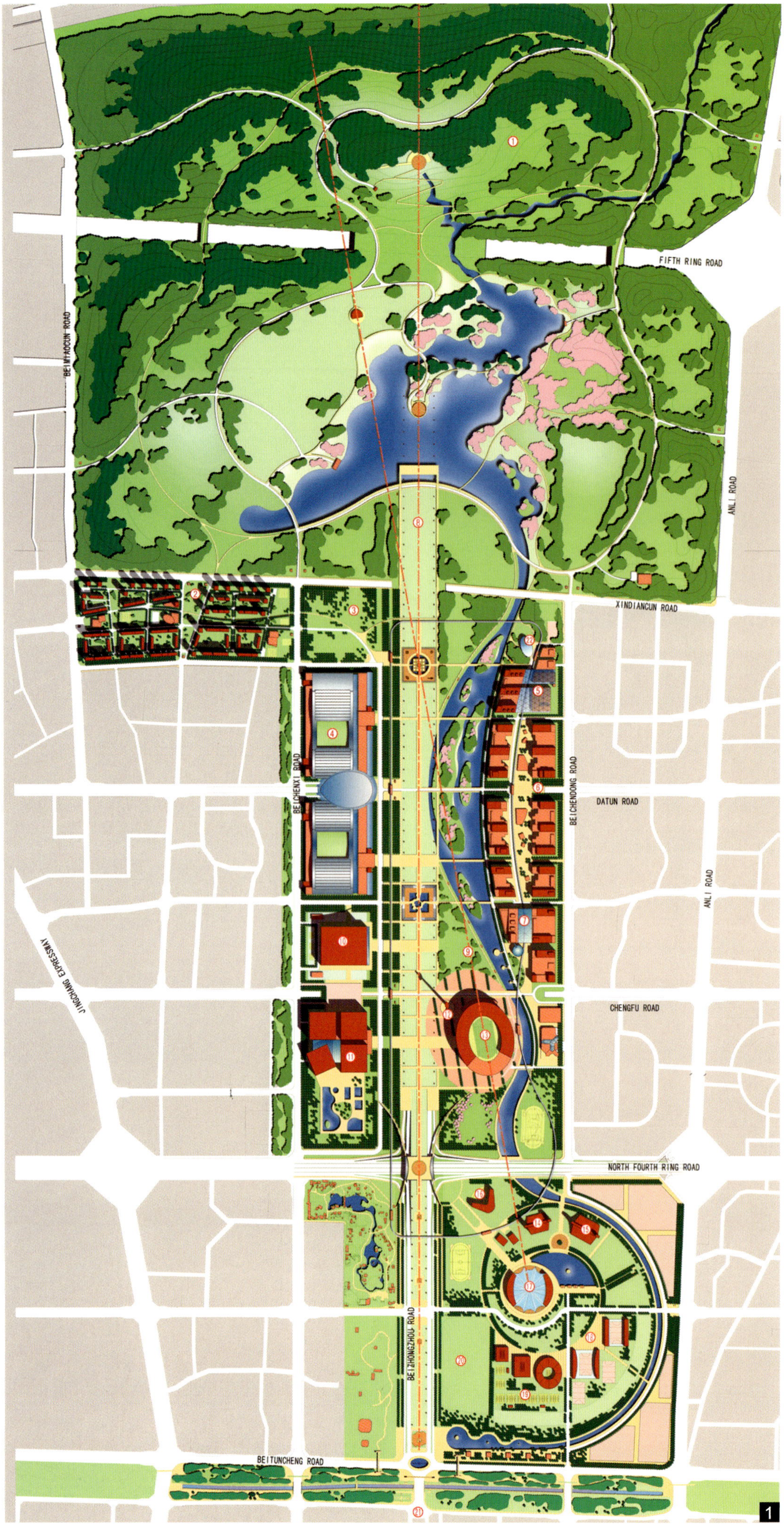

FIFTH RING ROAD

BEIYIAOCUN ROAD

ANLI ROAD

XINDIANCUN ROAD

BEICHENXI ROAD

BEICHENDONG ROAD

DATUN ROAD

ANLI ROAD

JINGCHANG EXPRESSWAY

CHENGFU ROAD

NORTH FOURTH RING ROAD

BEIZHONGZHOU ROAD

BEITUNCHENG ROAD

1

1. Master Plan
   总平面图

2. View of Canal Park with the
   National Stadium to the South
   国家体育馆南侧公园路径

3. View of the Edge of Forest Park
   with a View of Beijing
   北京森林公园周边

4. View of the Main Axis in Thirty
   Years
   30年的主奥运轴线

Vanke Experience Center

万科体验中心

Vanke Experience Center is located at the east exhibition hall of the Vanke Architecture Research Center at the No. 63 Meilin Road, Futian District, Shenzhen. With high design becoming an increasingly integral part of Vanke's brand and business model, the company wanted a place to show off the innovative architecture of its latest properties and the work of its research group. The designers responded to the client's brief by creating a curvaceous, three-story exhibition structure which takes full account of the sense of "experience" and "interest" of the indoor exhibition space. The fluid form and see-through metal-mesh skin prevent any sense of claustrophobia and inject a sense of playfulness inside the four-story more serious concrete research center. Additionally, it also drove the architects to treat the project as a public sculpture within an airy, park-like setting. The see-through metal-mesh skin, blurred boundary of the different areas, fluid space as well as the vigorous interior display make up this fabulous space.

万科体验中心项目，位于深圳市福田区梅林路63号万科建筑研究中心东侧展厅内，拟建三层展示空间，是一个室内改造项目。万科体验中心是为了协助完成生活设计研究项目而成立的，是万科生活研究设计组的"产品测试基地"。用于测试生活设计研究项目中所研发的创新产品的使用状态，同时，它也将为万科的研发部门与客户提供一个交流和"共同完成设计"的场所。在展示空间的设计上，甲方提出设计应充分考虑展厅室内空间在使用时的"体验性"和展示的"趣味性"。设计首先以现状场地条件为出发点，考虑作为一家房地产公司的研究展示建筑，万科体验中心展现在人们眼前的形象应该是鲜活生动，充满生活气息的。而它的基地现状却是一个四层通高，由混凝土柱支持起的面积逾1500平方米的大空间展厅，边界硬朗，气氛凝重。为了缓解这一矛盾，设计决定采用柔和的曲线来创造空间，从而改变场地的气质。与一般室内改造项目不同的是，万科体验中心在展示其内部功能的同时，也在对外展示形象(这是由其场地的透明性质决定的)。由此概念决定了建筑不但具备一定的完整性，同时又能体现其体验中心性质的信息。综上所述，便形成了现在的万科体验中心：金属网拼合起来的三维曲面表皮；柔性边界的平面；流动的内部空间，以及生动有趣的室内展示。

Project name: Vanke Experience Center
Award date: 2008
Location: Shenzhen, China
Building area: 2,600 m²
Architect: Urbanus Architecture & Design Inc.
Client: China Vanke Co., Ltd.
Completion date: 2006
Award name: McGraw-Hill Construction 2nd Bi-Annual "Good Design Is Good Business" China Awards 2008, Best Interior Project

项目名称：万科体验中心
获奖时间：2008
项目位置：中国 深圳
建筑面积：2，600平方米
建筑设计：都市实践
业主：中国万科有限公司
完成时间：2006
所获奖项：麦格劳-希尔公司《建筑实录》、《商业周刊》第二届"好设计创造好效益"中国奖项 2008最佳室内设计

1. Second Floor
   二层

2. Second Floor
   二层

3. Stairs
   楼梯

4. Stairs
   楼梯

1

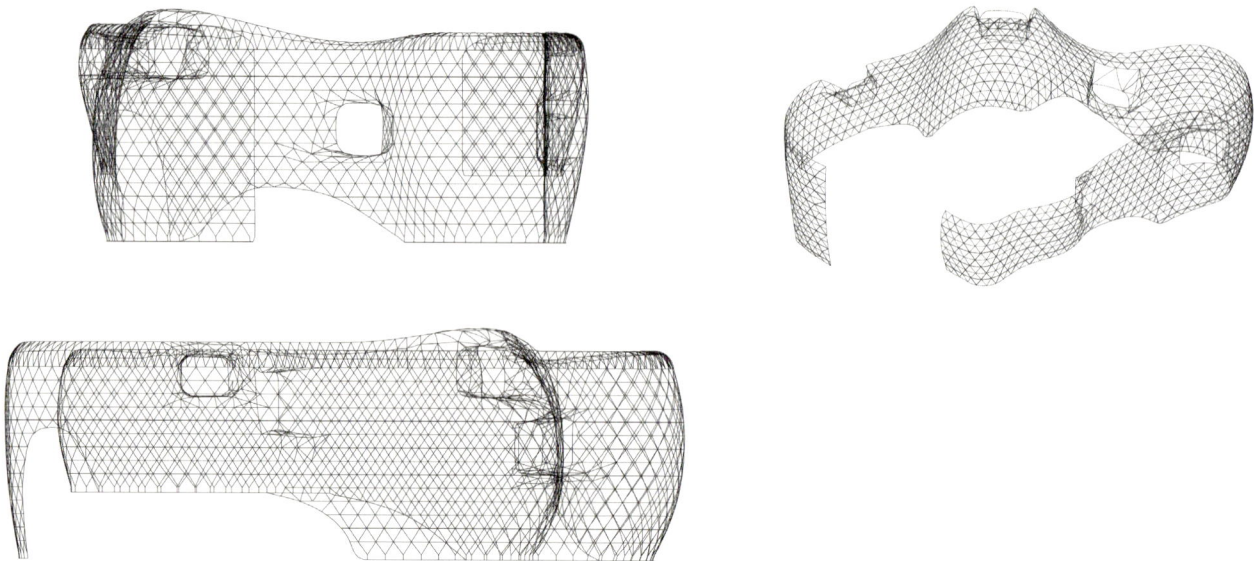

2

1. Detail
   细部

2. Stucture
   结构图

3. Entrance Hall
   入口大厅

4. Detail
   细部

Index 索引